TABLE OF CONTENTS

Cosmogenic nuclides

Anthropogenic radionuclides

Materials sciences

Education

EDITORIAL

It is a pleasure for us to summarize the activities of the **Laboratory of Ion Beam Physics** (LIP) and present to you an overview of the achievements of the laboratory accomplished in 2015. This annual report will provide you with a comprehensive and easy to read snapshot of all our research activities, in a format that became well established over the past years. We do hope you enjoy reading it.

The ongoing instrumental development process with radiocarbon dating equipment has reached an advanced level of perfection and it was possible to built three new MICADAS instruments in 2015: one was set into operation at the the commissariat a l'energie atomique et aux energies alternatieves, in Gif-sur-Yvette, France, a second one at the University of Bristol in the UK. The third instrument has been installed at LIP so that now also our own research activities profit from the latest stage of radiocarbon detection systems. The latter has been possible because of the generous financial support from the ETHZ Department of Earth Sciences which will be a major user of this instrument. All these instruments are equipped with the "GreenMagnet" technology and are utilizing He as stripper gas. They are the last MICADAS instruments which have been built by our laboratory. From 2016 onwards, the production of MICADAS became fully commercialized and will be executed by the Ionplus AG, an ETHZ spinoff company that was founded in 2013 to make instrumentation originally developed at LIP commercially available worldwide.

The transfer of knowledge and duties from LIP to Ionplus AG goes in parallel with a drain of technical expertise and man-power, because also the technical staff responsible for the production of MICADAS instruments and related components moved to the spinoff company. This is certainly a challenge for the operation of our lab and also for our capacity for developing advanced AMS instrumentations such as a multi-isotope version of the MICADAS instrument or the development of an even simpler radiocarbon system. However, LIP maintains a technical infrastructure which still represents an ideally suited platform for conducting a large variety of application projects. This versatile instrumentation, which has been improved by the installation of the new LIP MICADAS system, will continue providing both, excellent service to our internal and external users and significant contributions to the educational program of ETH.

Our applied research fields cover the wide range from fundamental research, over operational issues of the laboratory, to the vast variety of exciting applications of our measurement technologies. We conduct these studies not only in connection with our partners at Paul Scherrer Institut, Empa, Eawag, and other ETH departments, but also with our external collaborators from Swiss, European, and overseas Universities, from national and international research and governmental organizations as well as with commercial companies. We want to thank all our internal and external partners for their confidence and support.

Of course we are grateful to all LIP staff members. They have contributed diligently, with commitment, and with remarkable passion to the LIP activities. Without such an excellent scientific, technical and administrative staff, the success of the laboratory would not have been possible.

Hans-Arno Synal and Marcus Christl

THE TANDEM AMS FACILITY

Operation of the 6MV Tandem Accelerator

Remote controlled beam reduction grid

High intensity isobar separation

Energy straggling in gases - The high energies

OPERATION OF THE 6 MV TANDEM ACCELERATOR

Beam time statistics

Scientific and technical staff, Laboratory of Ion Beam Physics

In 2015, the 6 MV tandem accelerator was in operation for 1242 hours, similar to the previous year (Fig. 1). Again, most of the time was devoted to actual measurements or new developments, however about 11% of the time was used for conditioning and other accelerator related running times. We had five tank openings in 2015 because of a failure of the drive belt of the LE chain, a broken suppressor cable, a broken resistor in the corona regulation circuit, worn bearings on the LE drive side and a chain breakage at the HE side. During the repair of the corona regulation we also replaced the corona needles that draw current from the terminal to regulate the high voltage (Fig. 2).

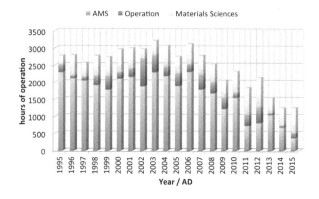

Fig. 1: *Time statistics of the TANDEM operation subdivided into AMS (blue), materials sciences and MeV-SIMS (green), and service and maintenance activities (red).*

The AMS activities concentrated on measurements of ^{36}Cl (173 samples) and ^{26}Al (63 samples) and developments at the gas-filled magnet. June 26th 2015 marks an historic date for the Tandem with the last measurement using the data acquisition (DAQ) system HAMSTER, which had been in operation since 1996. Failure of the VME computer hard drive and processor finally forced us to put the HAMSTER into a well-earned retirement.

Subsequently we built up a new DAQ based on the FASTComTec MCA3 system and the in-house development SQUIRREL including new readout of the Faraday cups. During the upgrade numerous old signal cables were removed from the racks in the control room. First AMS measurements with the new system were made in November 2015.

Fig. 2: *Corona mushroom with new needles.*

About one third of the time was used for developing the new MeV-SIMS setup CHIMP where first successful measurements with Au beams of several MeV could be performed.

For materials science applications the beam time decreased from 510 to 390 hours, mainly due to the long accelerator downtime. Still, approximately 1500 samples were analyzed by IBA techniques or irradiated. This is well above the long term average but below the peak value of more than 2000 samples in 2014.

After a long break of many years we have performed again activation runs with high energy proton and deuteron beams for two external groups. For radiation safety reasons these runs took place during evening hours and the whole accelerator complex was locked.

REMOTE CONTROLLED BEAM REDUCTION GRID

Optional remote beam intensity reduction and DAQ rate control

M. George, R. Gruber

Some of the measurements performed at LIP require tunable particle rates. After the beam setup and start-up is finished, the rates are potentially too high e.g. for RBS or MeV SIMS measurements.

Thus, a manually rotatable grid was installed in the beamline some years ago. The variety of different experiments and the consequential large set of different measurement conditions required regular manual intervention to adjust the grid rotation. Besides, the installed grid only covered transmissions between 15% and 2%.

To enlarge flexibility in rate reduction, transmission values of a set of new grids was tested (Fig. 1). Out of the tested grids, model number 6 showed a wide transmission coverage between ~45% and ~3% and consequently was selected for replacing the old reduction grid.

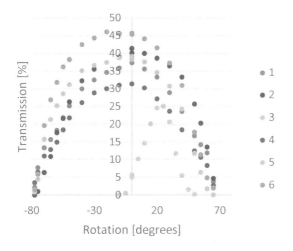

Fig. 1: *Transmission vs grid rotation for six different grid types.*

With the replacement process of the grid itself, a new support structure which has a stepping-motor connected to the axis was built. The whole structure can be pushed into the beam, when a reduction is necessary. This gives the possibility to control the beam intensity over a wide range remotely. The user interface for the grid control was written in LabVIEW (Fig. 2). The desired transmission (in %) can be selected through a slider. Based on a function that was fit to the transmission values shown in Fig. 1, the transmission selected through the slider is translated to motor steps.

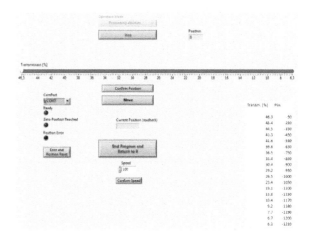

Fig. 2: *Remote control interface for beam reduction grid.*

As a further advantage, the grid control software can be linked to e.g. the RBS DAQ tool. In case a set of samples with very different properties, resulting in very different DAQ rates, is measured the beam transmission can be adapted to provide similar DAQ rates.

In the last step of the grid replacement, grid number 3 was mounted to a non-rotatable aperture, which can be placed into the beam additionally. This extends the covered beam transmission rate from ~45% down to ~0.2%.

HIGH INTENSITY ISOBAR SEPARATION

Reviving the Gas-Filled Magnet for ^{26}Al and ^{32}Si measurements

C. Vockenhuber, K.-U. Miltenberger, M. Suter

AMS measurements of several nuclides are challenging because of intense interference from isobars which cannot be directly resolved in the detector. Thus a significant reduction of the intensity of the isobar by several orders of magnitude is required before they can be identified in the ionization detector. One promising method is the gas-filled magnet (GFM) which separates isobars according to the different mean charge of the ions as they travel through the gas-filled magnet chamber. However, a relatively high ion energy of >1 MeV/u is required to achieve sufficient separation and keep the losses due to scattering under control. At the 6 MV EN Tandem a GFM was installed more than 20 years ago with promising first results. However, lifetime limitations of the then used plastic entrance foils prevented a use for routine measurements and no GFM measurements have been performed for many years. Now, with more robust silicon nitride (SiN) foils this drawback is gone – moreover they can be made thinner and are very homogeneous, reducing the effect of the entrance window on the final resolution.

In 2014 we performed first tests for ^{26}Al-^{26}Mg separation with injection of ^{26}Al^{16}O$^-$ to improve the overall efficiency [1], but we still had substantial losses (>80 %) because the beam after the GFM was too wide for the small (6×12 mm^2) detector window. In 2015 we installed a larger (30×40 mm^2) Mylar detector window to increase the beam acceptance and could measure the ^{26}Al standard ZAL94N to ≈55% of its nominal value of ^{26}Al/^{27}Al = 480×10^{-12} (Fig. 1).

Additional tests with ^{32}Si and ^{36}Cl showed also promising results. In the case of ^{32}Si the interfering isobar ^{32}S reaches the GFM with an intensity in the nA range. With the proper setting of the GFM the ^{32}S intensity is reduced to <1000 cts/s, allowing the separation of ^{32}Si and ^{32}S in the ionization detector (Fig. 2).

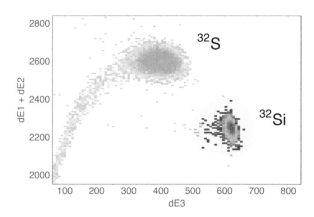

Fig. 2: *Separation of ^{32}Si and ^{32}S in the ionization detector after the GFM.*

A drawback of the present setup is leakage of detector gas through the Mylar foil into the GFM, resulting in drifts and changes of the ion energy. Furthermore, the used detector is too small for the large entrance window. A new detector is currently being designed and built for optimal performance of the GFM system.

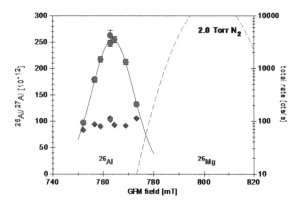

Fig. 1: *Separation of ^{26}Al from ZAL94N (blue symbols) and ^{26}Mg (red dashed line) in the GFM. The remaining ^{26}Mg counts (red symbols) are well separated in the ionization detector.*

[1] C. Vockenhuber et al., LIP Annual Rep. (2014)

ENERGY STRAGGLING IN GASES - THE HIGH ENERGIES

Complementary measurements at ETH EN Tandem and JYFL

C. Vockenhuber, J. Jensen[1], K. Arstila[2], J. Julin[2], H. Kettunen[2], M.I. Laitinen[2], M. Rossi[2], T. Sajavaara[2], H.J. Whitlow[3], P. Sigmund[4], A. Schinner[5]

In 2013 we performed a series of energy-loss measurements of ^{28}Si beam in various gases (He, N$_2$, Ne, Ar and Kr) at the 6 MV EN Tandem [1], where the energy range of 0.5-2.35 MeV/u was covered. In 2014 and 2015 we expanded the dataset to energies up to 12 MeV/u using the K130 cyclotron at the Accelerator Laboratory of the University of Jyväskylä (JYFL), Finland. In contrast to the measurements at ETH, where we used a windowless gas cell with a magnetic spectrometer, we measured the beam energy profile after a gas cell with Si$_3$N$_4$ windows with a double TOF setup (Fig. 1).

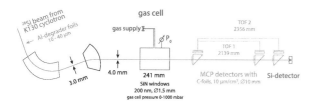

Fig. 1: *Experimental setup at JFYL.*

At each energy, data at several pressures were taken to confirm the linear dependence of straggling (variance Ω^2 of the energy-loss distribution) vs. gas pressure (Fig 2).

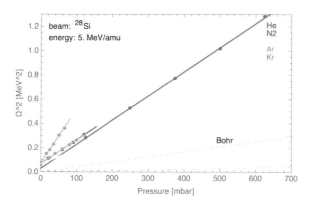

Fig. 2: *Straggling vs. gas pressure at 5 MeV/u. Data (points connected with solid line) and Bohr straggling (dashed lines) are shown.*

The ratio of the slope to the one calculated for Bohr straggling (Ω^2/Ω_{Bohr}^2) is shown in Fig. 3. Excellent agreement of the results at overlapping energies (at 1-2 MeV/u) from the measurements at the ETH EN Tandem and at JYFL (from beam times in 2014 and 2015) could be obtained.

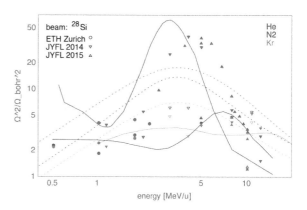

Fig. 3: *Results for He, N$_2$ and Kr; solid lines are the theoretical curves by Sigmund et al. [2] and dashed lines are empirical Yang formula [3].*

New theoretical calculations by Sigmund et al. [2] show a pronounced structure in the energy dependence (Fig. 3). The predicted peak for He gas could be confirmed, although the position and height is slightly off. Further measurements are planned to resolve the structure in more detail.

[1]　M. Thöni et al., LIP Annual Report (2013)

[2]　P. Sigmund et al., NIM B 338 (2014) 101

[3]　Q. Yang et al., NIM B 61 (1991) 149

[1] *Linköping University, Sweden*

[2] *University of Jyväskylä, Finland*

[3] *HE-Arc Ingénierie, La Chaux-de-Fonds*

[4] *Univ. of Southern Denmark, Odense, Denmark*

[5] *Johannes Kepler University, Linz, Austria*

THE TANDY AMS FACILITY

Activities on the 0.6 MV Tandy in 2015

A new standard for low ^{129}I samples

Improved measurements of ^{26}Al

ACTIVITIES ON THE 0.6 MV TANDY IN 2015

Beam time and sample statistics

Scientific and technical staff, Laboratory of Ion Beam Physics

In 2015, the multi-isotope facility TANDY (Fig. 1) accumulated more than 2700 operation hours.

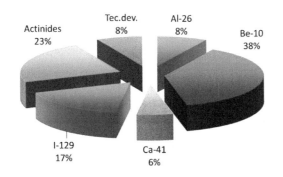

Fig. 1: *The multi isotope system TANDY.*

About 8% of the beam time was dedicated to testing and developing new instrumentation (Fig. 2), while the remaining time (92%) was spent for routine AMS analyses. During this time about 2000 unknown AMS samples were analyzed for different radionuclides and various applications (Fig. 3). The numbers impressively demonstrate that, over the past few years, the multi isotope system TANDY became the working horse for all non-^{14}C AMS applications at ETH Zurich.

Fig. 2: *Relative distribution of the TANDY operation time for the different radionuclides and activities in 2015.*

More than ½ of the AMS samples were analyzed for ^{10}Be, most of them for in-situ dating applications and ice core studies. About ⅕ (each) were measured for ^{129}I and the actinides involving a broad field of applications like environmental monitoring (e.g. Fukushima), human bioassay studies, and Oceanography (e.g. in the Arctic Ocean). Currently, large surveys are carried out to map and understand the distribution of ^{129}I and ^{236}U in the oceans. Therefore, we expect that the number of AMS samples for these nuclides will increase.

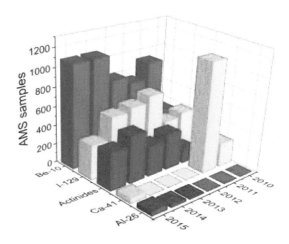

Fig. 3: *Number of AMS samples per nuclide measured over the past 6 years.*

In 2015, we also started with routine analyses of ^{26}Al, applying a novel absorber technique that allows the efficient detection of ^{26}Al in the 2+ charge state. Since this method now has reached application status we expect that the numbers of ^{26}Al samples will rise in 2016. Finally, a few samples were analyzed for ^{41}Ca as part of a pilot study for biomedical applications. Since the main part of this study will be carried out in 2016, also the number of ^{41}Ca analyses will increase in 2016.

A NEW STANDARD FOR LOW ^{129}I SAMPLES

Calibration of the low-level ^{129}I/^{127}I standard E1

C. Vockenhuber, M. Christl, N. Casacuberta

The 500 kV TANDY facility is well suited for AMS measurements of ^{129}I [1]. Since the stable isobar ^{129}Xe does not form negative ions high ion energies are not required for discrimination in the final detector. Moreover at low ion energies (terminal voltage of 300 kV) the transmission through the accelerator for charge state 2+ is high (>50 %); together with the high ionization yield in the ion source very efficient measurements are possible. Background issues are also well controlled by (a) excellent separation against the stable isotope ^{127}I is provided by the spectrometer at the high energy side; (b) molecular interferences (molecules with mass 129 and charge state 2+) are sufficiently reduced by increasing the stripper gas pressure, and (c) the mass to charge ratio (m/q) is not an integer number, reducing the chance of molecular breakup-products to reach the final detector.

The main limitation for ^{129}I measurements with the TANDY currently is the cross-contamination in our NEC MC-SNICS ion source. The volatile nature of iodine and the positioning of the samples on a 40 position wheel that is inside the ion source volume and close to the hot ionizer increases the risk of cross-talk, so that samples with high ^{129}I/^{127}I ratios can contaminate samples with low ratios.

In order to reduce the cross-talk between the samples care must be taken to which samples are loaded in the wheel. Usually low samples (with ^{129}I/^{127}I < 10^{-12}) are measured first while high samples are measured at the end of a measurement series with reduced Cs sputter intensity. For low samples our usual in-house standard D22 (^{129}I/^{127}I = 50.35×10^{-12}) [2] leads to severe cross contamination. In this case our low-level Woodward (WW) iodine blanks are measured around 1-2×10^{-13} which is a factor 5-10 higher than the expected value.

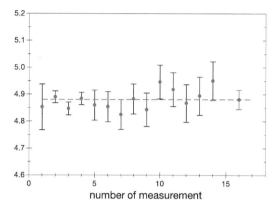

Fig. 1: *Individual measurements (blue) of the new E1 standard with the mean in red.*

The standard material E1 from the original dilution series has an order of magnitude lower isotopic ratio and thus reduces the risk of cross-talk for the low ratio samples. Over the past two years we carefully calibrated the E1 against D22 in five independent measurement series (Fig. 1). The new calibrated ratio is (4.88±0.10)×10^{-12} based on the mean of recent measurements. Older measurements of the E1 are not taken into account, but generally agree with the new calibrated ratio. Previous measurements had much larger uncertainties (typically 3-4%) due to the limited counting statistics in measurements at higher charge states and low transmission.

With the new established E1 standard it is now possible to measure low samples (e.g. form the deep Arctic Ocean [3]) without cross-talk from the high D22 standard and the measured ratios of the WW blanks are back in the 10^{-14} range.

[1]   C. Vockenhuber et al., NIM B 361 (2015) 445
[2]   M. Christl et al., NIM B 294 (2013) 29
[3]   N. Casacuberta et al., LIP Annual Report (2015) 79

IMPROVED MEASUREMENTS OF ^{26}Al

^{26}Al standard comparison and introduction of low ZAL02 standard

K.-U. Miltenberger, M. Christl, P. W. Kubik, A. M. Müller, C. Vockenhuber

For aluminium ions in charge state 2+ the use of helium as a stripper gas in the accelerator of the ETH 500 kV AMS facility TANDY enables a high transmission of more than 50 %. However, for measurements of ^{26}Al the intense interference of ^{13}C^{1+} entering the detector has to be suppressed. This could be achieved using a newly developed absorber setup for low energy ^{26}Al^{2+} measurements [1].

To prove the applicability of this absorber-detector configuration for ^{26}Al measurements using the 2+ charge state, the Nishiizumi standards [2] and real samples were measured and normalized to the ETH Zurich ZAL94N standard with a nominal ratio of $(480\pm18)\times10^{-12}$ [3]. Fig. 1 shows the measured to nominal ratios of the Nishiizumi standards and their overall weighted mean.

$(1–3)\times10^{-14}$ and are currently limited by ^{26}Al cross-contamination caused by the high ratio of the ZAL94N standard in the NEC SNICS ion source.

Therefore, the new ETH Zurich standard ZAL02 with a lower nominal ^{26}Al/^{27}Al ratio was introduced, to minimize cross-contamination during the measurement of samples with low ratios. The ZAL02 standard has been independently measured several times over the last years and calibrated against the Nishiizumi standard KN01-4-2 with a nominal ratio of $(30.96\pm1.11)\times10^{-12}$ [2]. All results from measurements at the Tandem and the TANDY agree very well. Again, the new low-energy setup at the TANDY achieves much smaller uncertainties.

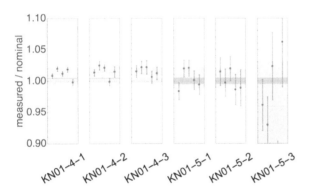

Fig. 1: *Measured to nominal ^{26}Al/^{27}Al ratios of the Nishiizumi standards normalized to ZAL94N (blue) and error associated with the ZAL94N nominal ratio (gray).*

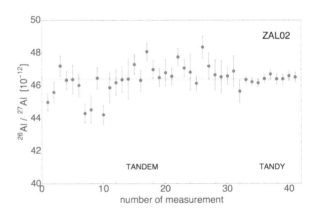

Fig. 2: *Individual measurements of the ETH Zurich standard ZAL02 and overall mean.*

The measured ^{26}Al/^{27}Al ratios agree well with earlier measurements conducted at the ETH 6 MV TANDEM AMS facility, but the statistical measurement errors are reduced significantly due to better counting statistics and more stable measurement conditions. Ratios of ^{26}Al blank material were measured in the range of

Based on the weighted mean of these measurements as shown in Fig. 2, the new calibrated ratio of the ETH Zurich ZAL02 standard is $(46.4\pm0.1)\times10^{-12}$. Not included in this value is the uncertainty of the primary KN01-4-2 standard (3.6 %).

[1]   A. M. Müller et al., NIM B 361 (2015) 257

[2]   K. Nishiizumi, NIM B 223-224 (2004) 388

[3]   M. Christl et al., NIM B 294 (2013) 29

THE MICADAS AMS FACILITY

Radiocarbon measurements on MICADAS in 2015

The years of MICADAS development

Status of the 300 KV MICADAS accelerator

RADIOCARBON MEASUREMENTS ON MICADAS IN 2015

Performance and sample statistics

Scientific and technical staff, Laboratory of Ion Beam Physics

In 2015, we broke the barrier of 10,000 analysed sample targets, an increase of 20% over the previous year. As the present runtime of 90% of the MICADAS cannot be increased anymore, this was only possible by improving measurement efficiency.

The EA-AMS coupling runs now in routine operation and allows measuring samples more efficiently, which resulted in 2000 additionally measured gas samples! Thus, with 5220 specimens we analysed more gas samples than the 4950 solid graphite samples (see also Fig. 1). For example, bulk sediment samples are no longer graphitised, but are directly analysed as gas samples. Additionally, we have developed the so-called *Speed Dating* for fast pre-dating of wood samples [1].

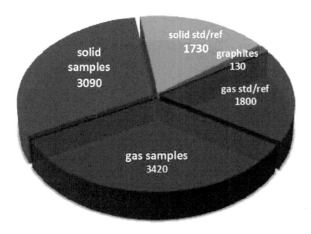

Fig. 1: *Samples measured on MICADAS in 2015. Graphite samples and standards prepared at ETH Zurich are in blue, samples graphitized outside ETH are in gray. Red indicates samples measured with the gas ion source.*

About 10% more samples (2700) than in the previous year were measured for our partner institutions. The samples analysed for internal projects was even doubled to 1500, while the ones for commercial projects stayed constant at 2000.

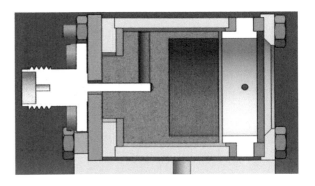

Fig. 2: *New cup design with a suppressor-ring at -300 V (white) at the entrance of the charge-collecting cup (green).*

Besides the time used to run the large amount of samples, there was not much room left to improve the MICADAS. Nevertheless, new cups were successfully tested for the measurement of $^{13}C^+$ and $^{12}C^+$. The previously installed cups were relatively long (20 cm) to minimize steric loss of backscattered electrons when positive ions are collected in the cups. The newly developed cups (Fig. 2) are in contrast much shorter (7 cm), but equipped with a suppressor ring at a voltage of -300 V at the entrance of the cup. Free electrons formed at the ion impact cannot escape the cup anymore and thus allow very accurate and stable measurement of positively charged ions.

[1] A. Sookdeo et al., LIP Annual Report (2015) 30

THE YEARS OF MICADAS DEVELOPMENT

Instruments now ready for commercial production at Ionplus AG

H.-A. Synal, S. Fahrni, M. Ruff, T. Schulze-König, M. Seiler, M. Stocker, M. Suter, L. Wacker

First experiments with a vacuum insulated high voltage platform to replace conventional accelerator technology in AMS instrumentation began in 2000, shortly after the successful demonstration that the 1+ charge state can be used to efficiently destroy molecules in multiple ion gas collisions. After a first proof-of-principle experiment performed with spectrometer components of the ETHZ Tandy AMS system in 2002, a prototype instrument named MIni CArbon DAting System (MICADAS) was set in operation in 2004. Of course being constantly improved, this system is until now the backbone of the ^{14}C measurements at ETHZ. In 2008, a system dedicated to biomedical radiocarbon AMS analyses was developed in collaboration with Vitalea Science, and installed at Davis, California. Based on this design a high-performance dating MICADAS was built and installed at the Reiss-Engelhorn Museum in Mannheim, Germany.

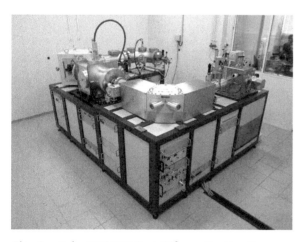

Fig. 1: *EchoMICADAS at Gif-sur-Yvette, France.*

The next generation of MICADAS systems incorporates the hybrid gas ion source, and several interface systems were developed to enable direct analyses of gaseous CO_2 samples. Related instruments are in operation in Seville, Debrecen, Brussels, and Berne. By replacing the

N_2 stripper gas with He, another performance upgrade was realized resulting in high ion beam transmission and related to this, an enhanced stability of the measurement conditions. The first system was installed at Aix-en-Provence in 2014. As a last development step, permanent magnets were introduced into the MICADAS setup. This reduces the electrical power to operate the instruments significantly and makes heat dissipation by water cooling obsolete. The first system was installed at Uppsala in 2014, and more recently in Gif-sur-Yvette, and Bristol (Fig. 1, 2).

Fig. 2: *MICADAS at University of Bristol, UK.*

The last MICADAS system built at ETHZ was completed in December 2015 and will be used by LIP to increase ^{14}C measurement capacity, in particular for ultra-high precision dating measurements. With the integration of He stripping and the implementation of permanent magnet technology, MICADAS systems have reached a development stage, which allows commercialization of production of further instruments. Consequently, ETHZ has ceased MICADAS system development projects, and outsourced further MICADAS activities to the ETHZ-spin-off company Ionplus AG [1].

[1] Ionplus AG, Dietikon, Switzerland

STATUS OF THE 300 KV MICADAS ACCELERATOR

Operational difficulties and technical improvements

S. Maxeiner, A. Herrmann, M. Christl, A. Müller, M. Suter, C. Vockenhuber, H.-A. Synal

A novel type of compact multi isotope AMS system is developed at LIP. It is based on a vacuum insulated accelerator also used in the MICADAS radiocarbon AMS systems which use acceleration voltages of up to 200 kV. To enable optimal measurements of a wide range of isotopes, acceleration voltages of up to 300 kV are desirable. Using the original MICADAS configuration though, a maximum of 260 kV could be applied to the terminal in an isolated test setup, before electric breakdown in the helium stripper gas feeding system destroyed the capillary (Fig. 1). In a subsequent test without gas feeding the full 300 kV could be applied, but after some days of operation the ceramic high voltage feedthrough broke down (Fig. 2).

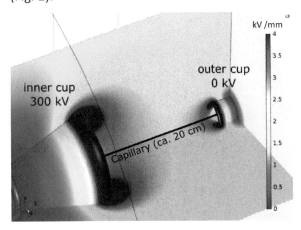

Fig. 1: *Electrostatics simulation of the helium gas feeding system showing the magnitude of the electric field gradient. Both fittings of the capillary are shielded from electric fields by cups, which by themselves introduce regions of higher gradients.*

With electrostatics simulations (using COMSOL) such as shown in Fig. 1, special cups were designed which still shield the capillary fittings from electric fields, but at the same time move regions of high field gradients away from the capillary containing the stripper gas.

The reason for the breakdown of the high voltage feedthrough (seen as a blackened hole in Fig. 2) was suspected to be a slight bending of the high voltage cable inside of the ceramic cylinder. This was improved by introducing special spacer rings, similar to a technique used at PSI. Furthermore, the feedthrough was carefully redesigned with the help of electrostatics simulations to lower field gradients at critical positions.

Fig. 2: *The ceramic of the high voltage feedthrough broke down due to excessive electric field gradients at a terminal voltage of 300 kV.*

After these improvements were implemented, the accelerator could be conditioned to the full 300 kV while feeding helium gas to the terminal. The voltage could be successfully held for several days, suggesting that the unit is now ready for prolonged operation at the full 300 kV acceleration voltage required for AMS measurements.

DETECTION AND ANALYSIS

Fully digital time of flight data acquisition

New RBS digital data acquisition

High voltage capillary electrophoresis

FULLY DIGITAL TIME OF FLIGHT DATA ACQUISITION

An infinite STOP Time-to-Digital-Converter

M. George, M. Schulte-Borchers, A.M. Müller

For the new MeV SIMS setup at LIP [1] a fully digital data acquisition approach has been implemented. Given the experiment's characteristics of having one STOP detector and up to three different START detectors, combined with the required time resolution on the order of few ns, the CAEN DT5751 digitizer (Fig. 1) was the most suitable choice. With all four channels active and equipped with pulse shape discrimination (PSD) firmware, it provides a sampling rate of 1 GS/s at 10-bit resolution.

Fig. 1: *CAEN DT5751 signal digitizer, 1 GS/s sampling rate, 4 channels, 10-bit resolution [1].*

Since the detector parameters for routine operation of the MeV SIMS setup are supposed to remain unchanged, a fixed set of acquisition parameters was determined for each detector. A LabVIEW based acquisition tool was developed (Fig. 2), which controls the digitizer operation and saves the relevant data (particularly timestamps) to disk. In a second step, this tool was extended. As additional functionalities dedicated data taking modes for high rate and low rate measurements were implemented, e.g. online trigger rate displays over time. In addition, a multi-stop time of flight analysis was implemented, which saves data to temporary files and from there calculates the final spectra online. The detection of multi-stop events is necessary, because a single incoming particle can create multiple secondary particles. Heavy particles have a longer flight time

through the spectrometer, than light particles. Thus, the correct spectrum can only be identified in case all secondary particles are assigned to the corresponding primary particle. Coincidence rates between the detectors are displayed for the active channels. All analysis steps are updated every 10 s.

After each run the full set of information, containing raw data to summarized DAQ monitoring parameters and final time of flight histograms, is saved to disk.

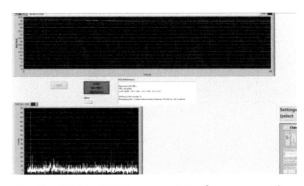

Fig. 2: *MeV SIMS user interface, providing digitizer control and data analysis; here: only one START detector is active.*

Stable operation was tested for trigger rates up to 18 kHz per channel in "low rate mode", which includes a spot check display of full waveforms. In "high rate mode" the waveform display is deactivated, which allows operation at trigger rates of up to 40 kHz.

In summary, the digital data acquisition has been put online for the new MeV SIMS setup and provides live information and analyses on the measurement.

[1] M. Schulte-Borchers, LIP Annual Report (2015) 89

[2] www.caen.it

NEW RBS DIGITAL DATA ACQUISITION

High Resolution RBS measurements with CAEN digitizers

M. George, A.M. Müller, M. Döbeli

At the beginning of 2015 the RBS setup was revised, including installation of a new Si PIN diode, a Cremat CR-110 preamplifier and the transition to routine operation with a signal digitizer. An evaluation and testing period led to the result that for RBS measurements the CAEN DT5780 digitizer (Fig. 1) is the most suitable model.

Fig. 1: *CAEN DT5780 digitizer, 100 MS/s sampling rate, 14-bit resolution [1]*

The digitizer is controlled via a custom made LabVIEW application (based on [2]), which also provides a live data display. In order to achieve highest resolution, a set of test measurements was done to optimize the pulse height analysis parameters to the new detector and preamplifier. Different sets of best-resolution acquisition parameters were created. Depending on the measurement conditions (high rate, low rate, high energy, low energy), the optimized and predefined settings can be loaded from config-files.

In comparison to the previous digital acquisition system, the software has been redesigned in order to cover high rate operation. Stable data taking up to 20 kHz was realized, which is beyond common operation conditions. Tests with a pulser showed the capability to acquire at higher rates. During qualification measurements it turned out that usage of a custom-made HV and LV power supply reduces noise on the signal lines and thus increases the resolution. Hence, the package of DT5780 digitizer and external

supply was put into routine operation for RBS measurements.

Altogether, parameter optimization studies allowed reaching a measurement resolution of 13 keV, e.g. visible in the 2 MeV He spectrum of a thin $Au_{0.2}Cu_{1.0}$ film on a Silicon substrate in figure 2.

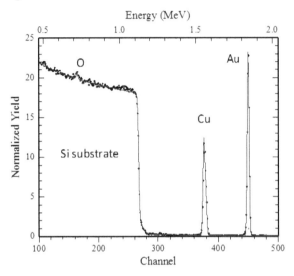

Fig. 2: *2 MeV He RBS spectrum of a thin $Au_{0.2}Cu_{1.0}$ film on a Silicon substrate.*

For the near future, intensive tests at high rates are planned. The aim of these measurements is to understand the hardware and software limits of high precision operation and to investigate the dependency of energy resolution versus rate. In case the measurement precision can be kept on a constant level at higher rates, the measurement time could be further reduced in the future.

[1]   www.caen.it
[2]   M. George, A.M. Müller, M. Döbeli, LIP Annual Report (2014) 32

HIGH VOLTAGE CAPILLARY ELECTROPHORESIS

High throughput separation technique for complex component mixture

S. J. Lee[1], C. Vockenhuber, H.-A. Synal, A. Manz[1]

Separation of complex component mixtures is very important for figuring out the morphology of certain kinds of cell structures, unknown chemical mixtures, and biomarkers for human diseases etc. Here we propose a high throughput separation technique called capillary electrophoresis (CE) [1] with a high voltage power supply. In a first experiment we used the conventional CE process for sample injection and separation. The target sample, four amino acids (4AAs, Tab. 1) with labeled FITC, was injected using a borax buffer as a background electrolyte and a detection system based on a laser induced fluorescence (LIF) system. The separation efficiency in CE is an important factor to characterize the separation power. Separation efficiency should increase proportionally to the applied voltage. To test this we used a 300 kV high voltage power supply for CE (Fig. 1).

	N	RSD (%)
GLN	7.42E+05	1.7
Ala	5.47E+05	1.7
Lys	5.85E+05	1.7
Glu	5.96E+05	1.8

Tab. 1: *Separation efficiency (N) and Relative standard deviation of migration time (RSD, %) for 4AAs separation with CE at 100 kV.*

The 4AAs sample was injected at 5 kV for 3 s and subsequently separated at 100 kV. Then, the electropherogram (Fig. 2) was collected and each peak was analyzed to calculate the separation efficiency. The achieved separation efficiency was approximately 750000 which is almost two times higher than commercial CE separation for amino acid separation. In future experiments (collaboration between KIST-Europe and LIP) we will try to increase the separation efficiency for CE at even higher voltages.

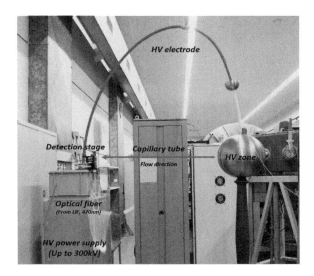

Fig. 1: *Experimental setup for CE with the 300 kV high voltage power supply. A special HV setup was built to hold up to 300 kV in air. The thin capillary tube follows the orange arrow to the detection stage on ground.*

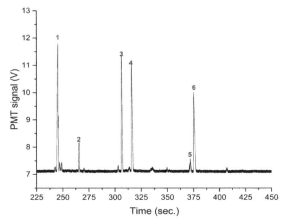

Fig. 2: *Electropherogram for 4AAs. at 100kV 1,2: free-FITC 3: GLN 4:Ala 5:Lys 6:Glu.*

[1] R. Weinberger, Practical Capillary Electrophoresis (2000) 462

[1] *Microfluidics, KIST-Europe, Saarbrücken, Germany*

RADIOCARBON

^{14}C Preparation laboratory in 2015

Myth Morgarten

How far can we get?

Speed dating

Growth constraints on a modern stromatolite

Different ^{14}C ages for different fractions of peat

Lake level reconstruction of Lake Sils, Engadine

Radiocarbon dating in the Falémé Project

^{14}C ages for the "Gonja Project"

Medieval capital of Cherven towns (Poland)

Gródek – mysterious Cherven town (Poland)

Indirect radiocarbon dating of a Pearl Oyster

Can red snapper live for half a century?

Investigating the organic carbon cycle in caves

Particle flux processes in the Arctic Ocean

Organic carbon cycling in Taiwan

Time-series ^{14}C studies of ocean particles

^{14}C age of thermal organic decomposition

Pool-specific radiocarbon measurements in soils

^{14}C PREPARATION LABORATORY IN 2015

Overview of samples prepared for ^{14}C analysis

I. Hajdas, S. Fahrni, M. Maurer, C. McIntyre, M. Roldan, A. Synal, L. Wacker

All samples, which are submitted for ^{14}C AMS dating, require selection of fractions that will result in accurate age determination of the sample. This is then followed by pre-treatment and conversion to graphite or CO_2. The development in activities of the ETH preparation laboratory is summarized by the number of samples that were prepared for different types of applications during the year 2015, as compared with previous years (Fig. 1, Tab. 1).

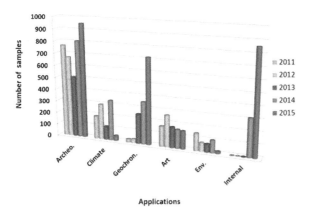

Fig. 1: Number of samples (objects) analysed for various research disciplines during the last five years.

Research	Total	Internal
Archaeology	963	12
Past Climate	41	
Geochronology	1560	844
Art	149	
Environment	18	
Total	**2731**	**856**

Tab. 1: Number of samples analysed in 2015 for various applications. Column 'Internal' shows the number of samples prepared as a part of laboratory research and in the frame of master or term theses.

The last 2 years showed an increasing number of research projects that are shown as 'Internal'

(Fig. 1). Such projects, which are leading to a better understanding and improvement of preparation methods and radiocarbon time scale, are an important part of our laboratory development. Moreover, many 'Internal' samples are an integral part of thesis (BA, MA, term papers) or school projects. In such studies samples are often analyzed after various preparations or different fractions are compared. This is reflected in numerous targets that were prepared and analyzed in 2015.

For the past 5 years 'Archaeology' remained the main application with a slightly growing number. The growing share of 'Geochronology' reflects a shift from Climate and Environmental research to interdisciplinary projects. This is also visible in the material showing growing numbers of wood and foraminifera samples (Fig. 2). The number of samples in category of 'Art' remained stable during the last 3 years. It is important to note that most of the samples shown here were prepared as graphite samples, however in the case of low carbon content samples are measured as CO_2 using GIS.

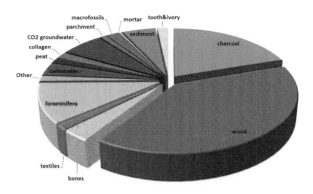

Fig. 2: Type of material prepared and analysed at the ETH laboratory in 2015.

MYTH MORGARTEN

ETHZ dated the bolt from Hünenberg

H.-A. Synal, I. Hajdas, L. Wacker, A. Fischli Roth[1], T. Müller[1], J. Frey[2], R. Hugener[3]

To mark the 700th anniversary of the "Battle of Morgarten" possible artifacts from this event have been investigated on behalf of "Einstein", the science program of the Swiss Television SRF. One of the most exciting stories entwines around the "Hünenberger Pfeil" (Fig. 1). According to the myth, Heinrich von Hünenberg should have shot this arrow through the fortifications at Arth, to warn the troops of the Swiss Confederation, and disclose the attack plan of the impeding Habsburg army of knights. The arrow is owned by the corporation Unterallmeind Arth and is on public display together with a testimony of the Zay family issued by the Registry of the Canton of Schwyz in 1862, which certifies the oral tradition of its authenticity.

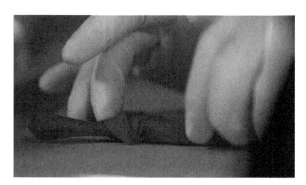

Fig. 1:   *Head of the crossbow bolt.*

Technically seen the object is a medieval crossbow bolt. An archaeological survey showed that the arrow head may be dated in the second half of the 14th century. To verify this a ^{14}C analysis of the wooden shank of the bolt has been carried out at ETH Zurich Laboratory of Ion Beam Physics. For this purpose, two samples were taken from the center of the shaft, processed independently and analyzed at the MICADAS. The resulting radiocarbon age of 588±10 years BP needs to be calibrated to obtain true historic age ranges corresponding to the growth of the wood from which the bolt was

made. With 95% probability these are the intervals from 1314 to 1357 AD and from 1388 to 1405 AD (Fig. 2). The historical date of the Battle of Morgarten (Nov. 15, 1315) thus lies within these time limits. Is it therefore the real "Hünenberger Pfeil"?

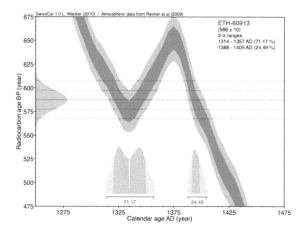

Fig. 2:   *Calibration of the result of the ^{14}C test.*

Just the outermost ring of a tree represents the date when it was cut down while the inner parts are older according to the number of annual rings. The arrow shaft consists out of several annual rings that do not originate from the outermost part of the trunk. Thus, one can assume that the cutting date is indefinitely younger than the historic date of the battle. The typological dating of the arrow head (2nd half of the 14th century) is impressively confirmed by the results of the radiocarbon analysis. Accordingly, the legendary crossbow shot, if it truly happened, took place unlikely with this arrow. However, the literary myth may continue to live on.

---

[1] *Schweizer Radio und Fernsehen*
[2] *Amt für Städtebau Zürich*
[3] *Staatsarchiv des Kantons Zürich*

HOW FAR CAN WE GET?

1‰ radiocarbon measurements on a single cathode

H.-A. Synal, S. Fahrni, A. Sookdeo, L. Wacker, D. Galvan[1]

Over the last ten years, significant progress in radiocarbon AMS at our MICADAS instruments has been realized. The use of He instead of N_2 as stripper gas has improved performance with respect to overall detection efficiency, and the optimized ion optical transport results in very reproducible measurement conditions. The recent implementation of permanent magnets reduces complexity of the instruments, significantly reduces operation and installation costs, and improves measurement stability.

Year AD	Year BP	σ/Years
770	1322	9
771	1299	9
772	1331	9
773	1317	9
774	1293	9
775	1235	9
776	1199	9
777	1196	9
778	1171	9
779	1185	9
780	1187	9

Tab. 1: *Results with 1 σ overall uncertainties.*

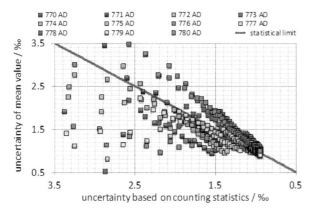

Fig. 1: *Uncertainties of individual cathodes as function of measurement time.*

As a result of these developments ^{14}C analyses can now be performed with total uncertainties of only ~1‰ on a single cathode. Here, we report on measurements on annual tree ring samples from 780-770 AD, a time range with unexpected atmospheric ^{14}C variations as observed by Miyake et al. [1]. In Fig. 1 reproducibility of the mean values of repeated measurements of individual samples with increasing counting statistic is demonstrated. All individual samples approach nicely the Poisson statistics limit at more than 1 million ^{14}C counts per sample. 3 Oxa I and 4 Oxa II standards were measured resulting in an uncertainty of the normalization procedure of less than 0.5‰.

Phthalic Anhydride, brown coal, and Kauri wood samples were used to assess blank levels. In addition, two pine wood samples cut from a tree ring grown in 1515 AD were nicely reproduced with ^{14}C ages of 343±8 BP and 348±8 BP, respectively.

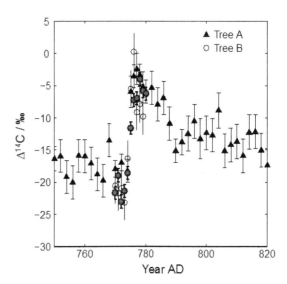

Fig. 2: *Comparison with data from Miyake [1].*

[1] F. Miyake et al., Nature 486 (2012) 240

[1] *Swiss Federal Research Institute WSL*

SPEED DATING

A rapid way to determine the radiocarbon age of wood by EA-AMS

A. Sookdeo, L. Wacker, S. Fahrni, C.P. McIntyre, M. Friedrich[1,2], F. Reinig[3], B. Kromer[2], U. Büntgen[3], W. Tegel[4]

Trees are discovered in construction sites, lake sediments and rivers (Fig.1) but the ages of the trees are not known until a dendrochronology link, i.e. the analysis of tree ring growth patterns to trees of known age, is established or if a radiocarbon (^{14}C) date is determined, both of which are time consuming and are expensive. If the trees are not in time periods of interest or in cases where dendrochronology links or ^{14}C dates are not lacking, the effort to date these trees is frivolous. In addition, the authenticity of historical buildings or sites can be called into question and dendrochronology or radiocarbon dates can be used to prove or disprove the age of the building or site, which again is a large and expensive undertaking [1]. At the Laboratory for Ion Beam Physics (LIP) we developed a rapid and cheap method to determine a rough ^{14}C date for wood material, using an Elemental Analyzer (EA) coupled with AMS. These ^{14}C dates can be used to determine whether or not a site warrants further investigation and aid in placing trees in dendrochronology series. We call this method Speed Dating.

Until present, we have dated at LIP over 600 samples found in France, Switzerland, Germany, and Italy. Samples speed dated from Germany have allowed for the potential extension of the Preboreal Pine Chronology (PPC) [2] to 14,500 cal. BP (Fig.2).

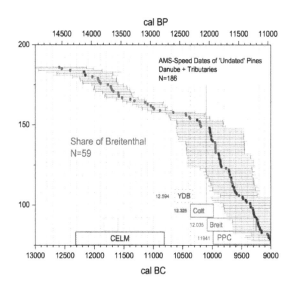

Fig. 2: *Pines that have been placed in a dendrochronology series with the aid of speed dating.*

[1]   B. Dietre et al., Quat. Int. 353 (2014) 3
[2]   M. Fredrich et al., Radiocarbon 46 (2004) 1111

[1] *Environmental Physics, Heidelberg University, Germany*
[2] *Botany, Hohenheim University, Stuttgart, Germany*
[3] *Swiss Federal Research Institute, WSL, Birmensdorf*
[4] *Archaeology, Environmental Science, Forestry, University of Freiburg, Germany*

Fig. 1: *Oak found at the bottom of lake Aigle, Switzerland.*

GROWTH CONSTRAINTS ON A MODERN STROMATOLITE

Stromatolite laminae as archives of reservoir age changes (S. Atlantic)

S. Bruggmann[1], I. Hajdas, C. Vasconcelos[1]

As laminated biogenic or abiogenic sedimentary structures [1], stromatolites record environmental changes along growth profiles, possibly revealing changes in reservoir ages. A modern stromatolite sample (Fig. 1) was collected in Lagoa Vermelha (100 km east of Rio de Janeiro, Brasil), an area known for upwelling of South Atlantic Central Water (SACW).

Fig. 1: *Lagoa Vermelha, arrow indicating north.*

Hand-drilled carbonate samples from different layers were analyzed for ^{14}C content. Recently collected shells were used to estimate the present-day reservoir age. The OxCal depositional model (Marine13 calibration curve; [2]) was used to calibrate ^{14}C ages. The maximum age in the center of the stromatolite was in the 6th century AD and the outer crust grew in the beginning of the 20th century. The well-laminated middle part of the stromatolite transect was found to have grown in a short time period of less than 100 years, with four excursions towards older ^{14}C ages. To detect the causes of these changes of marine ^{14}C, calendar years assuming a stable modern reservoir age were used to simulate atmospheric ^{14}C ages with the southern hemisphere ShCal13 atmospheric calibration curve [3]. The offset between the measured and simulated ^{14}C ages

indicates a variability of the reservoir age between -99 and 268 ^{14}C y. The highest reservoir corrections correlate with layers indicating environmental changes as detected by other geochemical measurements. These analyses support an increase in the intensity of upwelling, bringing old deep water enriched in ^{14}C to the lagoon.

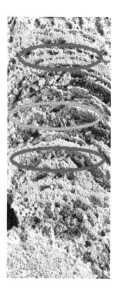

Fig. 2: *Circles highlight three layers with excursions towards older ^{14}C ages caused by upwelling.*

This additional old carbon causing the observed excursion are very well recorded in stromatolites, making them excellent archives for reconstructing reservoir ages.

[1] M. Semikhatov et al., Can. J. Earth Scie. 19 (1979) 992

[2] P. Reimer et al., Radiocarbon 55 (2013) 1869

[3] A. Hogg et al., Radiocarbon 55 (2013) 1

[1] Geology, ETH Zurich

DIFFERENT ^{14}C AGES FOR DIFFERENT FRACTIONS OF PEAT

Insights from a depth-profile in a fluvial terrace of Ticino River

J. Braakhekke, I. Hajdas, G. Monegato[1], F. Gianotti[2], M. Christl, S. Ivy-Ochs

The glacial landforms in the lower valleys of northern Italy have been assigned to different ice ages by a variety of authors since the 19th century [1]. This study is part of a project that involves a broad geomorphological analysis and first-time absolute in-situ exposure dating of erratic boulders using ^{10}Be and ^{36}Cl.

Seven radiocarbon samples were taken from a fluvial terrace outcropping over a height of 6 meters along the Ticino River (Fig. 1 and 2).

Fig. 1: *Hill shaded LIDAR image of the sample location on the left bank of the Ticino River.*

Where possible, the samples were sieved to separate a bulk fraction (<125 µm) from the undefined organic fragments. During the subsequent ABA-preparation for all fractions some samples dissolved partly. This way we obtained up to four ages per initial sample; one each for the insoluble bulk fraction, the humic acid of the bulk, the organic fragments and the humic acid of the organic fragments.

The radiocarbon ages vary significantly, with 20 ky between the insoluble bulk fraction and the organic fragments. All samples of the bulk fractions gave much younger ages than of the hand-selected macro remains (Fig. 2).

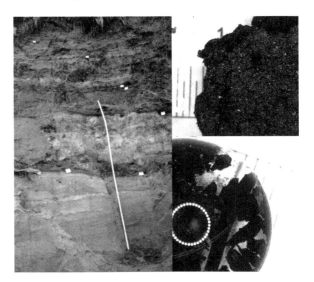

Fig. 2: *Outcrop with sample locations (left) and two microscopic pictures of a peat sample and selected undefined organic fragments.*

Age differences can be attributed to contamination of the sampled material with young rootlets, the passage of 'young' groundwater and very low carbon contents in especially the insoluble bulk. The ages of the organic fragments showed most coherent ages over the whole profile and these fragments are least likely contaminated by younger material.

Based on our results obtained on macro remains this deposit is dated to MIS3 age. About 3 meters of sediment have been deposited here during ca. 8 ky. This could tell us more about the sediment budgets of the interstadials and the erosive power of glaciers in the Last Glacial Maximum.

[1]   A. Penck and E. Brückner, 3 (1909) 1119

[1] *Geosciences and Earth Resources (CNR), Turin, Italy*
[2] *Earth Sciences, University of Turin, Italy*

LAKE LEVEL RECONSTRUCTION OF LAKE SILS, ENGADINE

Radiocarbon dating and dendrochronology of tree trunks

S. Vattioni[1], I. Hajdas, M. Strasser[1, 2], R. Grischott[1], T. Sormaz [3]

Lake level changes influence lacustrine and fluvial sedimentation processes and therefore also the appearance of landscape. In summer 2011, the Archaeological Service of Canton Grisons found several tree trunks in about 2.5 meters depth (fig. 1). It can be assumed that they come from trees that grew there when the lake level was lower than today.

Fig. 1: *Tree trunks found on the bottom of Lake Sils.*

A previous study has proposed that the lake level was significantly lower in the Middle Holocene and continuously rose between 3000 and 2000 years BP [1]. This scenario could have led to the death of the trees. Therefore, in this study we dated these tree trunks using radiocarbon dating and dendrochronology to test this scenario. The results revealed four different groups of ages of the trees. The first group of trees, which dates to the 8th century BC and the second one, which dates to Roman times indicate two lake level minima of about 3 meters below the present value. This lake level correlates with climate data such as Alpine glacier fluctuations [2]. Advancing glaciers could have transported more bed load to the fan delta of Sils, which is damming Lake Sils. Therefore, the lake level was subsequently higher resulting in flooding of the trees.

Fig. 2: *Map of Lake Sils showing the note about "Reusen".*

The two younger groups of trees with Medieval ages and the one with an early 20th century age, can be related to human activities. The trees from Medieval period were identified as piles of fish traps (German: "Reusen", Fig. 2). None of the proposed hypotheses could be confirmed. More likely, the lake level was about 3 meters lower than today until about 200 AD, followed by a lake level rise to about today's level, which was reached at about 300 AD [3]. This study, which combines dendrochronology with radiocarbon dating, shows high potential for the reconstruction of the Quaternary landscape history in the Alpine regions.

[1]   R. Grischott, PhD thesis, ETHZ (2015)
[2]   H. Holzhauser Holocene 15 (2005) 789
[3]   S. Wohlwend MS thesis, ETHZ (2010)

[1] *Geology, ETHZ*
[2] *Geology, University of Innsbruck, Austria*
[3] *Dendro Lab, Archaeological Service of Canton Grison, Chur*

RADIOCARBON DATING IN THE FALÉMÉ PROJECT

Chronology of Human Settlement and paleoenvironment in West Africa

I. Hajdas, M. Maurer, M. Roldan, E. Huysecom[1], B. Chevrier[1], S. Loukou[1], A. Mayor[1], members of Ounjougou Project[2]

Preliminary research conducted in the Falémé Valley, East of Senegal, have shown the potential of this region for allowing joint studies on both new Palaeolithic/Neolithic data and local palaeo environmental data [1]. Three main time periods are addressed: Palaeolithic, Protohistory and history of the last centuries. All archaeological and palaeo environmental studies on a section of the Falémé Valley (Fig. 1) involve geochronological investigations, which apply 2 techniques: OSL and ^{14}C.

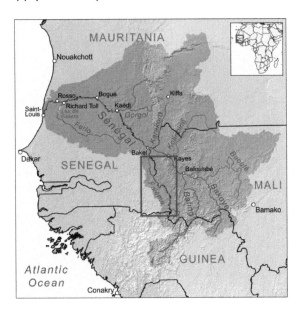

Fig. 1: Map of western Africa showing research area in the Falémé River Valley.

Each year a set of radiocarbon ages obtained on organic matter collected during the January-March field seasons provide a time frame for the sites surveyed and excavated. Typically, charcoal samples are collected in the field and sent to the laboratory for preparation and analysis. These are often of a very low size (Fig. 2) therefore careful monitoring of sample treatment is important. Nearly 50 samples were analyzed during the last 3 years were. The ages covered time from MIS2 to the most times.

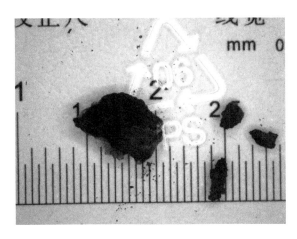

Fig. 2: Sample of charcoal as submitted to the laboratory i.e., prior to treatment. The resulting amount of pure carbon was 0.3 mg.

One of the examples is the dating of archaeological layer at Toumboura, which allowed to attribute its occupation to the end of the Pleistocene or early Holocene [2]. More recent stratigraphic sections might contain well preserved organic matter but here another challenge of the radiocarbon dating has to be acknowledged. Calibration of radiocarbon ages for samples from historic period of the last 2 or 3 centuries results in a wide range of calendar ages. However, in this project radiocarbon chronologies can be supported by stratigraphic information allowing Bayesian models of OxCal to be applied.

[1]   E. Huysecom et al., Journal of African Archaeology 13 (2015) 7
[2]   B. Chevrier et al., Quat. Int. (in press)

[1] Archaeology, University of Geneva
[2] http://www.ounjougou.org

^{14}C AGES FOR THE "GONJA PROJECT"

Preliminary results for 2015 excavation in Ghana (Northern Region)

D. Genequand [1], I. Hajdas, K. Gavua[2], W. Apoh [2], H. Amoroso [3], F. Maret [4], Ch. De Reynier [5]

The state or kingdom of Gonja emerged in the mid-16th century in the savannah area of Northern Ghana extending to the north of the tropical forest. According to historical and oral traditions, its origins go back to the arrival of Mande conquerors who came from the area of Djenne to take control of the gold trade from the mines situated around Buna and Beghu. They crossed the river Black Volta and established themselves in the area known as Gonja.

Fig. 1: Old Buipe, general view of the architectural remains in Field C at the end of the season (photo Denis Genequand).

The general aim of the project is the study of the islamisation of Northern Ghana from the 16th century onwards through the archaeological study of the Gonja. The detailed study of Old Buipe (Northern Region), one of the major archaeological sites of Gonja, includes topographic survey, soundings and extensive excavations (Fig. 1) to document the ancient phases of the town (16th to 18th century). Six samples of charcoal collected in the 2015 excavations confirm the ancient settlement.

Fig. 2: Bole, the mosque seen from the north (photo Denis Genequand).

In addition to studies on the origin of Gonja, a renewed inquiry into the architecture and the date of construction of some of the last surviving traditional mosques of Northern Ghana will be made as the majority of these monuments are situated in the territory of Gonja. Their origin and date of construction are still uncertain. First results of ^{14}C dating on samples of wood and charcoal collected at the Bole mosque (Fig. 2) confirm the construction after 1896-97 AD destruction of the town. These results are encouraging and essential for the upcoming excavations of 2016.

[1] *Laboratory Archaeology and Population in Africa, University of Geneva*
[2] *Department of Archaeology and Heritage Studies, University of Ghana*
[3] *Musée Romain d'Avenches, Avenches*
[4] *TERA SA, Sion*
[5] *Office of Heritage and Archaeology of the Canton of Neuchâtel (NAFO), Neuchâtel*

MEDIEVAL CAPITAL OF CHERVEN TOWNS (POLAND)

^{14}C chronology and palaeoenviroment of Czermno

R. Dobrowolski[1], I. Hajdas, M. Wołoszyn[2,3], J. Rodzik[1], I. A. Pidek[1], P. Mroczek[1], P. Zagórski[1], T. Dzieńkowski[1], K. Bałaga[1]

Early Medieval history of the territory of Hrubieszów Basin on the Bug river, that bordered Polish and Russian states at that times, has been investigated in terms of palaeoenvironment. The rich spectrum of issues associated with this region includes development of the so-called Cherven towns. One of the most recognized is the fort of Czermno. The stronghold was situated on a well-drained holm at the confluence of two rivers: Huczwa (tributary of the Bug) and its small tributary – Sieniocha. The fortifications were surrounded by a group of open settlements on the marshy bank of the Huczwa. The stronghold went out of use during the second half of the 13th century.

Fig. 1: *Archaeological excavation in the outer zone of the settlement complex.*

In addition to archaeological excavations (Fig. 1) environmental aspects of the location and operation of the fortified hill were investigated. The dense network of geo- and pedological drilling and precise topographical surveys were conducted. Geological cores and undisturbed samples were collected for pollen and ^{14}C analysis. These allowed reconstruction of an early medieval relief (Fig. 2) and of the original river network systems. The development phases of the town and its wide suburbia were also clarified.

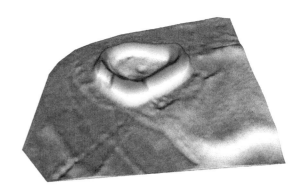

Fig. 2: *3D model of stronghold in Czermno.*

Samples of peat and organic (Fig. 3) had to be carefully sieved to remove modern rootlets. Our ^{14}C analysis and their Bayesian age model sets the timing of this first settlement phase in the early medieval.

Fig. 3: *Sequence of sediments with artifacts in the internal part of ramparts*

---

[1] *Geography, Maria Curie-Sklodowska University, Poland*
[2] *Archaeology, Rzeszów University, Poland*
[3] *GWZO, Leipzig, Germany*

GRÓDEK – MYSTERIOUS CHERVEN TOWN (POLAND)

Preliminary results of 2015 field work in Eastern Poland

R. Dobrowolski[1], I. Hajdas, M. Wołoszyn[2,3], J. Rodzik[1], I. A. Pidek[1], P. Mroczek[1], P. Zagórski[1], T. Dzieńkowski[1], K. Bałaga[1]

The small Early Medieval stronghold of Gródek (Hrubieszów Basin, Eastern Poland) was probably called Volyn at its beginnings, as can be deduced from historical documents. Gródek belongs to the Cherven towns region together with Czermno, neighbouring it from the south.

Fig. 1: *Location of stronghold at Gródek on map with main elements of relief.*

The settlement complex at Gródek, some 15 ha in area, consists of the remains of a fortified hill and a number of open settlements. The ramparts were lost in some places to artillery trenches dug during the two great wars of the 20th century. Archeological excavations proved the existence of a cemetery inside the hill fortifications, out of which over forty graves contained jewelry ornamented in the style of Kievan Rus of 12th- 14th century.

Fig. 2: *Present day view on the stronghold at Gródek.*

One of the aims of the present study was to reconstruct geologic-hydrologic conditions of the investigated strongholds and to reconstruct the palaeoenvironment in the settlement area.

Results of multidisciplinary studies of the ^{14}C dated peat core from the wetland on the river terrace of Huczwa (2 km from the stronghold) recorded two phases of intensified human impact on the area: end of 6th-8th century AD and end of 13th-14th century AD.

From the second half of the 10th century to the first half of the 12th century the anthropogenic effect of land cultivation is weaker coinciding with slightly drier climatic conditions.

[1] Geography, Maria Curie-Sklodowska University, Poland

[2] Archaeology, Rzeszów University, Poland

[3] GWZO, Leipzig, Germany

INDIRECT RADIOCARBON DATING OF A PEARL OYSTER

Rapid analysis of a marine carbonate sample by Laser Ablation AMS

C. Welte[1], L. Wacker, B. Hattendorf[1], M.Christl, A.H. Andrews[2], C. Yeman, H.-A. Synal, D. Günther[1]

Precise age estimates of the black-lipped pearl oyster are of great importance for fishery regulation and protection of this endangered species [1]. The radiocarbon (^{14}C) - record established from corals of the Kure Atoll can provide information about the life-span of the oysters. This approach requires to access the ^{14}C-signature along the growth axis of the shell at a high spatial resolution. The fastest method so far available for this is the novel Laser Ablation (LA)-AMS technique recently installed at the LIP [2]. The carbonate sample is placed in an in-house designed LA-cell, where CO_2 is produced from the sample by focusing a UV laser onto its surface. The CO_2 is directly transported into the gas ion source of the AMS and analyzed online. Two consecutive scans in opposite direction were performed along a line across the polished section of the oyster embedded in epoxy. In order to increase the measurement precision for each growth layer, a zig-zag scanning pattern was selected. The vertical displacement covered 2 mm, while the overall scanning distance was 12 mm. The spatial resolution achieved is 220 µm and the overall analysis time was 30 min. The unprocessed data are shown in Fig. 1.

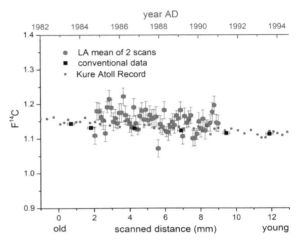

Fig. 2: *Data from LA (red), data from analysis on graphite targets (black) and Kure Atoll Record (purple).*

For the final LA-AMS data set, only regions, where both scans are available were considered. The first 2.5 mm of the scan from old to young were rejected, as parts of the epoxy were hit by the laser and caused a depleted F^{14}C signal. The average of the two scans was taken (red circles in Fig. 2). Furthermore, data from graphite analysis (black squares) are shown. The LA-AMS data matches the graphite data within the uncertainties. Nevertheless, the slope is slightly steeper, which cannot be explained at this point. The Kure Atoll record (purple squares) was aligned with the oyster data by considering the collection year of the oyster (1994). Furthermore, the slope of the Kure Atoll record was matched with the slope of the LA-AMS data, suggesting an age of the oyster on the order of ten years.

[1]   E. Keenan et al., Atoll Res. Bul. (2006) 543
[2]   C. Münsterer et al., Chimia 68 (2014) 215

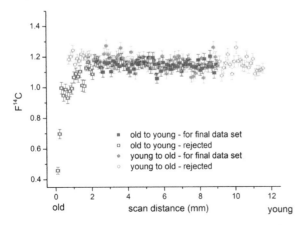

Fig. 1: *Raw data from two LA scans. Fraction modern is plotted versus position.*

[1] *D-CHAB, ETH Zurich*
[2] *NOAA Fisheries, Pacific Isl. Fisheries Sc. Center, USA*

CAN RED SNAPPER LIVE FOR HALF A CENTURY?

Laser-Ablation AMS reveals complete bomb ^{14}C signal in an otolith

A.H. Andrews[1] C. Welte, C. Yeman, L. Wacker, B. Hattendorf[2], D. Günther[2]

Red snapper (*Lutjanus campechanus*) is an important fishery species in the Gulf of Mexico. Fishery sustainability is supported by knowledge of valid life history parameters, like longevity. Estimates of maximum age for fishes were often underestimated because the method was not validated. For red snapper, otoliths (fish ear stones) are used to estimate age by counting growth zones in cross sections (Fig. 1).

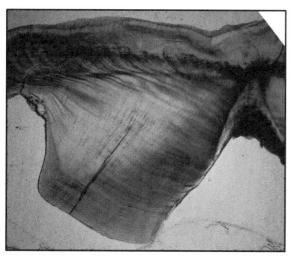

Fig. 1: *Cross-sectioned red snapper otolith aged to 50-55 years from growth zone counting.*

Recently collected red snapper were estimated to live more than 50 years, leading to birth years prior to atmospheric testing of thermonuclear bombs. Given the age estimate and birth year scenario is accurate, calcium carbonate accreted during the earliest growth (otolith core in upper right of Fig. 1) would have pre-bomb ^{14}C levels, followed by a rapid rise in ^{14}C, terminating in a post-bomb peak and subsequent decline (Fig. 2).

The age of fishes have been validated using traditional sampling of the otolith core (earliest growth) by extracting calcium carbonate with a micromilling machine and measuring ^{14}C for an alignment with a coral ^{14}C reference [1]. This method is typically limited to a single sample

because the temporal specificity of micromilling becomes difficult to impossible as the otolith grows in successive layers that become thinner with increasing age (Fig. 1). Hence, a fish with a pre-bomb birth year can be aged only to the coral ^{14}C reference inflection point at 1958 because ^{14}C levels plateau (Fig. 2).

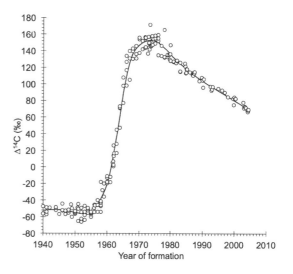

Fig. 2: *Bomb ^{14}C record for Gulf of Mexico from numerous coral records across the basin.*

A recent application of LA-AMS to a red snapper otolith section has led to a continuous series of ^{14}C measurements. This series has revealed pre-bomb ^{14}C levels in the earliest growth out to a point where the rapid ^{14}C increase is evident, providing a time-specific marker at a given fish age. This result provides an opportunity to age the fish with greater certainty and beyond the pre-bomb limitation at 1958.

[1] Andrews et al., Can. J. Fish. Aquat. Sci. 70 (2013) 1131

[1] *NOAA Fisheries, Pacific Islands Fisheries Science Center, Hawaii USA*
[2] *D-CHAB, ETH Zurich*

INVESTIGATING THE ORGANIC CARBON CYCLE IN CAVES

Extraction and measurement of ^{14}C from organic matter in stalagmites

F. Lechleitner[1,2], C. McIntyre[3], S. Lang[4], T. Eglinton[1], J. Baldini[2]

Stalagmites are valuable archives for past terrestrial climate. They can be dated using the U-Th method, yielding very precise chronologies. As a multi-proxy archive, they can provide a wealth of information on past rainfall distribution, temperature, vegetation and soil conditions. Small amounts of organic carbon entrapped in stalagmite carbonate are potentially useful indicators for past surface conditions. Drip water originating from the soil above the cave carries information about the soil and vegetation system in the form of organic carbon. Radiocarbon analysis of the organic carbon fraction could therefore give valuable insights into soil cycling and karst turnover times. However, the small concentrations of OC typically found in stalagmites (e.g. 1.73-8.8 µg lipid extract/g $CaCO_3$ [1]) present analytical challenges.

Fig. 1: *Schematic of a cave system and the associated hydrological and carbon cycle processes.*

We conducted a series of ^{14}C measurements on stalagmites from different locations. A wet

chemical oxidation method [2] was used in conjunction with acid digestion of the carbonate to measure the amount and ^{14}C-concentration of organic carbon in the stalagmites.

The oxidized samples were analysed on the MICADAS using a gas ion source device to directly sample headspace CO_2 in the sample vials.

Fig. 1: *Yok Balum cave, Belize, richly decorated with stalagmites, stalactites and flowstones. (Picture credit: I. Walczak)*

[1] A.J. Blyth et al., Quat. Science Rev. 27 (2008) 905

[2] S.Q. Lang et al., Limnology and Oceanography: Methods 11 (2013) 161

[1] *Geology, ETH Zurich*
[2] *Earth Sciences, Durham University, UK*
[3] *Scottish Universities Environmental Research Centre UK*
[4] *Earth and Ocean Sciences, University of South Carolina, USA*

PARTICLE FLUX PROCESSES IN THE ARCTIC OCEAN

A coupled organic and inorganic tracer approach

M.S. Schwab[1], J.D. Rickli[1], J. Blusztajn[2], S. Manganini[2], H.R. Harvey[3], A. Forest[4], R.W. Macdonald[5], D. Vance[1], C. McIntyre, T.I. Eglinton[1]

A growing body of evidence suggests that delivery of particulate matter, including associated biogeochemically relevant materials to the interior Canada Basin in the central Arctic Ocean, is dominated by lateral inputs. The magnitude and origin of lateral inputs has substantial implications for our understanding of biogeochemical cycling in the central Arctic Ocean and its impact on ecosystems as well as on records preserved in underlying sediments. In this study, organic (C/N, $\delta^{13}C$, $\Delta^{14}C$) and inorganic (Nd, Sr) tracers are combined to elucidate the sources of organic matter in core-top sediments and sediment traps of the North American Arctic Ocean including the Bering Sea.

Particulate organic carbon of pelagic and sea ice algae within arctic systems are depleted in ^{13}C and are thus similar to ^{13}C of vegetationally derived carbon, thereby complicating the deconvolution of distinct sources. Aged carbon, originating from the surrounding permafrost soils and kerogen-rich sediments delivered by rivers allows a distinction between marine and terrestrial inputs. Surface sediments represent a mixture of two end-member compositions only, modern marine organic matter and a combination of Mackenzie derived particulate organic matter and kerogen (Fig. 1). Young terrestrial carbon is of minor significance for Mackenzie Shelf sediments, which can be explained by two complementing factors, namely the remineralization within the water column and the aging of the carbon within the surface sediments. Samples of the Mackenzie and Colville Shelves approaching the compositional range of kerogen indicate the substantial input of petrogenic hydrocarbons along with aged organic matter of deep permafrost soils.

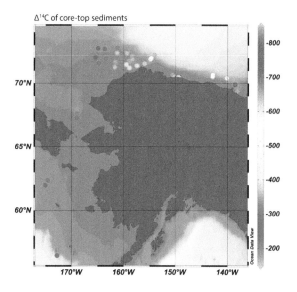

Fig. 1: *Compilation of radiocarbon data. Unpublished data from this study (n=32) and literature data (n=24)* [1,2,3].

Nd and Sr isotopic com-positions, allow the identification of continental sediment sources. The Bering-Chukchi Sea region is dominated by contributions from contemporary marine productivity, whereas the Beaufort Sea and the Canada Basin receive dominantly sedimentary inputs from the Chukchi Peninsula, the Alaskan Colville and the Canadian Mackenzie River.

[1] M. Goñi et al., Geophys. Res. 118 (2013) 4017

[2] K. Schreiner et al., Geophys. Res. 118 (2013) 808

[3] J. Vonk et al., Geochim. Cosmochim Acta 171 (2015) 100

[1] *Geology, ETH Zürich*
[2] *Woods Hole Oceanographic Inst., USA*
[3] *Old Dominion Uni., USA*
[4] *Universté Laval, Canada*
[5] *Department of Fisheries and Oceans, Canada*

ORGANIC CARBON CYCLING IN TAIWAN

Microbial oxidation of rock-derived organic carbon in tropical soils

J. D. Hemingway[1], R. G. Hilton[2], N. Hovius[3], T. I. Eglinton[4], V. V. Galy[1], C. McIntyre[4], N. Haghipour[4]

Oxidation of uplifted petrogenic organic carbon (OC_{petro}) is a net source of CO_2 to the atmosphere and counteracts drawdown by burial of biospheric organic carbon (OC_{bio}) in marine sediments [1]. This process is most apparent in erosive mountain systems exhibiting high OC_{petro} content, such as the Eastern Central Range (ECR) of Taiwan (Figure 1) [2]. A previous study has utilized dissolved rhenium concentrations in rivers draining the ECR to estimate that this process corresponds to a transfer of 12 – 20 tons $C/km^2/yr$ to the atmosphere [3]. However, a mechanistic understanding remains elusive.

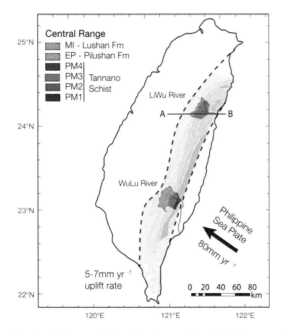

Fig. 1: *Map of the Eastern Central Range in Taiwan, emphasizing the sampling locations (LiWu and WuLu River catchments).*

Here we use stable carbon isotopes ($\delta^{13}C$), bulk radiocarbon (F_{mod}), vascular plant fatty acid F_{mod}, and OC thermal composition [4] as tracers in ECR soils and riverine suspended sediments to reveal that this process is microbially mediated within the soil column. Microbes rapidly and

efficiently utilize OC_{petro} as substrate upon uplift of fresh bedrock, leading to an estimated 7 – 38 tons $C/km^2/yr$ transfer to the atmosphere, consistent with previous estimates [3]. Additionally, bacterial fatty acid (FA) biomarker $\delta^{13}C$ values in a saprolite layer indicate that microbes incorporate OC_{petro} into radiocarbon-dead living biomass, as has been suggested previously using culture experiments [5].

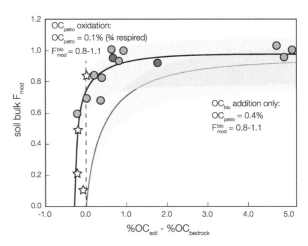

Fig. 2: *OC mixing model for Taiwanese soils. Addition of OC_{bio} without OC_{petro} (blue envelope) is inconsistent with the data, which instead suggest oxidation of ~75% of bedrock OC_{petro} (gray envelope).*

[1] V. Galy et al., Science 322 (2008) 943
[2] V. Galy et al., Nature 521 (2015) 204
[3] R. Hilton et al., Earth Planet. Sci. Lett. 403 (2014) 27
[4] B. Rosenheim et al., Geochem. Geophys. Geosys. 9 (2008) Q04005
[5] S. Petsch et al., Science 292 (2001) 1127

[1] *Marine Chemistry and Geochemistry, WHOI, USA*
[2] *Geography, Durham University, UK*
[4] *Geology, ETH Zurich*
[3] *Geomorphology, GFZ, Potsdam, Germany*

TIME-SERIES ^{14}C STUDIES OF OCEAN PARTICLES

Exploring land-ocean carbon isotope gradients in the South China Sea

T. Blattmann[1], T. Eglinton[1], Z. Liu[2], K. Wen[2], J. Li[2], Y. Zhao[2], Y. Zhang[2], L. Wacker, M. Plötze[3]

The South China Sea (SCS) borders the Pacific Ocean as one of its largest marginal seas. Sediment traps have been used to collect a time series of sinking particles in the water column in order to study sediment sources and fluxes in the SCS. The time series sediment traps (Fig. 1) were deployed on moorings at two different locations and at two water depths per mooring (1000-3000 meters depth). The traps include a carousel that enables discrete samples to be recovered each representing 18-day time-slices during the year-long mooring deployment. Intra-annual variability of particulate matter settling through the water column depends on the seasonality of marine primary productivity and continental discharge. In the case of the latter, special events such as eddies deliver huge amounts of sediments to the deep sea [1].

Fig. 1: *A sediment trap recovered from the South China Sea after one year of deployment. The samples are refrigerated and transported to the laboratory.*

Stable carbon and radiocarbon isotopic composition of bulk organic carbon was measured from the time series trap samples to gain insight into the provenance and type of organic matter exported from land and surface ocean waters to the deep sea. The radiocarbon isotopic composition shows subtle yet clear changes in the source material over the course of one year. During time windows when sediment fluxes were higher, radiocarbon concentrations are lower. These episodes likely reflect the export of aged terrestrial organic carbon superimposed on the pelagic background sedimentation. In the SCS, a significant part of this terrestrial component can be radiocarbon-dead organic carbon derived from bedrock erosion from Taiwan [2].

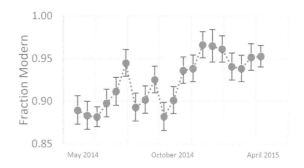

Fig. 2: *One year time series (from late spring 2014) of radiocarbon isotopic composition of bulk particulate organic carbon intercepted by a sediment trap deployed at 2000 m water depth in the South China Sea. Changes in the isotopic composition reflect changes in the type of organic matter transported to 2000 meters of water depth at this sediment trap.*

[1]   Y. Zhang et al., Sci. Rep. 4:5937 (2014) 1
[2]   R. Hilton et al., Geology 39 (2011) 71

^{1} *Geology, ETH Zurich*
^{2} *State Key Laboratory of Marine Geology, Tongji University, Shanghai, China*
^{3} *Geotechnical Engineering, ETH Zurich*

^{14}C AGE OF THERMAL ORGANIC DECOMPOSITION

Influence of lateral transport time on organic carbon degradation

R. Bao[1,2], M. Zhao[3], A. McNichol[2], N. Haghipour[1], C. McIntyre[1], T. Eglinton[1]

Organic matter (OM) on the continental shelf is subject to extensive degradation after laterally conveying and vertically sinking. Although previous studies proposed that transport-related oxygen exposure time correlated with OM preservation along the continental margin [1]. However, further studies are still required to quantitatively estimate the timescale of OM lateral transport and, in particular, to establish the relationship between lateral transport time and OC degradation.

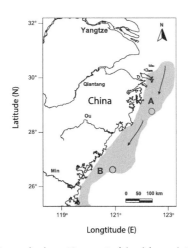

Fig. 1: Sample locations: A (dash) and B (circle) and dominant delivery direction (arrows) of marine sediments in the inner shelf of the East China Sea. Shadow area represents the mud region (modified from Xu et al., 2007)[2].

A new approach was recently developed for determination of carbon isotopic compositions of sedimentary OM using a so-called ramped pyrolysis/oxidation ("Ramped PyrOx") method [3]. This approach can provide radiocarbon analysis of the organic thermal decompositions, which are CO_2 that evolves under a linear temperature program, allowing separation of OC components in sediments based on their thermochemical stability.

Here we report the first attempt to explore the role of lateral transport time of OM associated with different grain size in OC degradation in the inner shelf of the East China Sea (ECS). We apply the Ramped PyrOx approach to measure the ^{14}C ages of the thermal decompositions of different grain size fractions from A and B surface sediments (Fig. 1). Our results show the squeezing of thermographic curves between A and B locations, suggesting OC degradation during lateral transport (Fig.2a). The ^{14}C ages of the thermal decompositions become older with increasing temperature on the OM, which in B thermal decompositions are systematically older than that of A corresponding decompositions (Fig. 2b). We propose that the lateral transport time of OM may be one of the most fundamental factors to influence OC degradation in the continental margin seas.

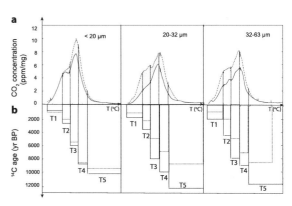

Fig. 2: (a): Thermographic patterns of OM from different grain size fractions in location A (dash) and B (solid); (b): ^{14}C age spectrums of thermal organic decomposition fractions (low-high temperature: T1-T5).

[1] R. G. Keil et al., Mar. Chem. (2004) 92
[2] K. Xu et al., Cont. Shelf Res. (2009) 29
[3] B. R. Rosenheim et al., G-cubed (2008) 9

---

[1] *Geology, ETH Zurich*
[2] *NOSAMS, Woods Hole, USA*
[3] *Ocean University of China, Qingdao, China*

POOL-SPECIFIC RADIOCARBON MEASUREMENTS IN SOILS

Comprehensive assessment of soil organic matter vulnerability

T. S. van der Voort[1], C. Zell[1], C. McIntyre, E. Graf Pannatier[2], F. Hagedorn[2], T. Eglinton[1]

Soil organic matter (SOM) forms the largest terrestrial pool of carbon outside of sedimentary rocks, and fulfills important ecosystem functions. Radiocarbon is a powerful tool for assessing SOM dynamics and is increasingly used in studies of carbon turnover in soils. Details on which specific fractions and compounds are responsible for SOM (in)stability remains convoluted when only measuring the bulk samples.

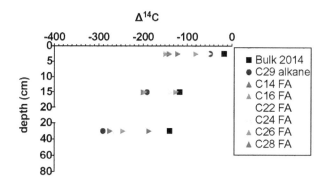

Fig. 2: *Compound-specific plant wax and fraction-specific radiocarbon analysis in Beatenberg.*

Initial results show that (1) Dissolved Organic Matter (DOM) signatures correlate with long-term and short-term temporal trends (Fig. 1) (2) density fractions reflect different SOM pools with largely variable turnover (3) compound-specific mineral-bound analysis reflect a larger range of ages than fractions and therefore may represent different SOM pools (Fig. 2). This, in combination with a bulk time-series of bulk organic carbon for four soil types, gives a uniquely comprehensive overview and understanding of soil organic matter dynamics. In conclusion, this work in progress reveals interesting trends in fraction- and compound-specific SOM pools that help to constrain pool-specific turnover. This work will be followed up with additional compound-specific work (lignin), turnover modeling and correlation to environmental parameters.

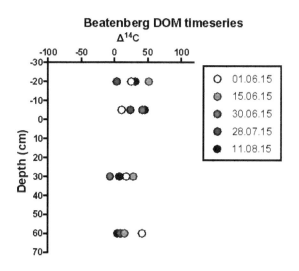

Fig. 1: *Radiocarbon analysis of Dissolved Organic Matter (DOM) in time series in LWF site Beatenberg, Switzerland.*

Therefore, this study also focuses on specific carbon fractions and compounds in order to trace specific pools within the SOM cycle. Overall, this is aimed to comprehensively assess the controls on organic matter stability and vulnerability in soils across Switzerland. Sites are part of the Long-term Forest Ecosystem (LWF) Research Program of the Swiss Federal Research Institute for Forest, Snow and Landscape Research (WSL).

[1] *Geology, ETH Zurich*
[2] *WSL, Birmensdorf*

COSMOGENIC NUCLIDES

Introducing the MECED Model

Glacial history in Patagonia

Latest glacier advances in the Tatra Mountains

Late Glacial landscape development in Meiental

Late Holocene evolution of the Triftjegletscher

The lateglacial in Rhaetian Alps

Reaching and abandoning the furthest ice extent

The last retreat of the Reuss Glacier

Late Pleistocene glaciers in northern Apennines

Synchronous Last Glacial Maximum

Timing of fluvial terrace formation in Europe

Past landscape change in the Hohen Tauern (A)

Past rock glacier activity in the northern Alps

Rock avalanches in the Mont Blanc Massif (Italy)

The mid-Holocene increased neotectoniv activity

Dating the Sentinel Rock avalanche of Zion, Utah

Dating Holocene basalts and human footprints

Isochron-burial dating: Terrace chronologies

Lithology conditions sediment flux

Denudation rates over space and time in Europe

Denudation rates from ^{10}Be (meteoric)/^{9}Be ratios

Constraining erosion rates in semi-arid regions

Erosion of the Central Bolivian Andes

Uplift rate distribution in the Andes (~32° S)

Climate control on Alpine hillslope erosion

Frost cracking as driver of Holocene erosion

The impact of typhoon Morakot on erosion rate

INTRODUCING THE MECED MODEL

A Multi-nuclide Exposure, Coverage and Erosion Depth profile model

C. Wirsig, S. Ivy-Ochs, V. Alfimov

During the course of our project [1] we developed a model that calculates depth profiles of cosmogenic nuclide concentrations in rock surfaces depending on pre-defined exposure scenarios. For each time period, the model calculates the production and decay of ^{10}Be, ^{14}C and ^{36}Cl taking into account a) exposure times, b) cover times, types and thickness and c) erosion rates. The type of cover not only defines its density: Fig. 1 shows the production rate depth profile of ^{36}Cl under sediment (green lines) or ice cover (blue lines), each equivalent to 16.6 cm of rock. As ice reflects thermal neutrons that would otherwise diffuse from the rock to the atmosphere, the ^{36}Cl production through this pathway is increased in the uppermost few decimeters of rock [2]. The magenta lines show the ^{36}Cl production that is 'lost' to the rock surface, because it occurs in the cover.

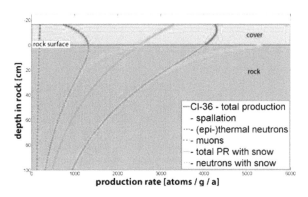

Fig. 1: *Production of ^{36}Cl under sediment (green lines) or ice cover (blue lines), each equivalent to 16.6 cm of rock. Calculated for sample Grub2 with (31.7 ± 0.5) ppm Cl [1].*

This model can be used 1) to conduct feasibility studies that test if expected (differences in) nuclide concentrations can be measured by AMS before investing long hours in the field and laboratory, and 2) to compare predicted final nuclide concentrations of diverse exposure scenarios to concentrations derived from AMS measurements of real rock surfaces. We can then restrict the set of exposure scenarios based on measured nuclide concentrations. In particular, the model predicts the concentrations of multiple cosmogenic isotopes and demonstrates the enhanced capabilities to constrain the duration of exposure and burial as well as the depth of erosion experienced by rock surfaces in studies using same-sample multiple isotope analyses.

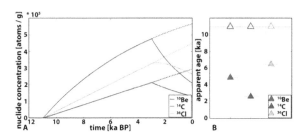

Fig. 2: A) *Evolution of nuclide concentrations starting at 11 ka with constant exposure. Solid lines with complete cover and 0.1 mm/a erosion in the last 3 ka.* **B)** *Apparent ages calculated from the final concentrations of A [1].*

Differences in the shape of the production rate depth profile and in the half-life of different nuclides cause characteristic patterns of apparent ages. For example, identical apparent ages indicate constant exposure, whereas cover and erosion are expected to result in apparent ^{36}Cl ages > ^{10}Be ages > ^{14}C ages (Fig. 2). Typical analytical uncertainties (grey areas) are low enough to allow the detection of an age mismatch between the different nuclides caused by the scenario in Fig. 2.

[1] C. Wirsig, Ph.D. thesis, ETH Zürich (2015)

[2] J. Fabryka-Martin, Ph.D. thesis, University of Arizona (1988)

GLACIAL HISTORY IN PATAGONIA

Geomorphological control on areal extent of ice

A. Cogez[1], C. Darvill[2], K. Norton[3], M. Christl, F. Herman[1]

The terminal moraines east of the Lago Buenos Aires (LBA) in Patagonia are particularly interesting because they are imbricated from the older (being around 1 Ma) to the east, to the younger, (from the Last Glacial Maximum, around 20 ka), to the west (Figs. 1 and 2). A precise chronology is however still lacking.

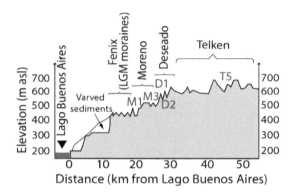

Fig. 1: Imbricated moraines east of LBA (adapted from [1]). M1, M3, D1, D2, T5 are the moraines for which we report an age (see fig. 3).

We sampled cobbles from the moraines outwashes and two depth profiles in Telken 5 and Deseado 2. We measured cosmogenic ^{10}Be concentrations to determine exposure ages.

Fig. 2: View over the imbricated moraines from the top of Moreno 1 moraine.

With these ages we establish a new chronology for the area: Moreno moraines, are from the same glacial stage (MIS 8). Deseado 1 is from MIS 12, Deseado 2 is from MIS 16, and Telken 5 probably from MIS 20 (Fig. 3).

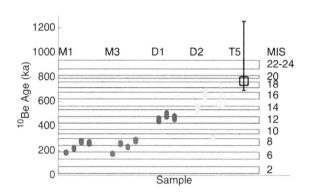

Fig. 3: Exposure ages determined for different cobbles (filled circles), and the two depth profiles (open square) (see fig. 1).

This chronology matches rather well with the one established by [2], in the Lago Pueyrredon area, which suggest that the processes controlling the areal extent of ice have a regional impact. One glacial cycle out of two is conserved, so that some moraines are recycled in the followings. Climate seems to play a minor role in glacier areal extent. We propose, in agreement with [3], that this pattern is explained by the morphology of the overdeepening (the lake) dug by the glacier.

[1] Singer et al., Geol. Soc.Am. Bull. 116 (2004) 434
[2] Hein et al., Quat. Sci. Rev. 29 (2010) 1212
[3] Kaplan et al., Geomorphology 103 (2009) 172

[1] *Geosciences, University of Lausanne*
[2] *Durham University, UK*
[3] *University of Wellington, New Zealand*

LATEST GLACIER ADVANCES IN THE TATRA MOUNTAINS

Insights from geomorphological mapping and surface exposure dating

E. Opyrchał[1], S. Ivy-Ochs, J. Zasadni[1], P. Kłapyta[2], C. Wirsig, G. Guidobaldi[3], S. Casale[3], M. Christl

During the Pleistocene, glaciers advanced several times in the Tatra Mountains. At the transition to the Holocene glaciers finally retreated from the massif, however, the exact timing is still under debate [1]. Therefore, the aim of this study is to determine when glaciers ultimately vanished from the Tatra Mountains using cosmogenic nuclides and relative dating methods. An additional goal is to obtain an age calibration curve which could be further applicable in similar mountain areas.

Fig. 1: *Location of the Rovienky Valley site (red square) in the Tatra Mountains.*

One of the most prominent and highest Lateglacial moraines within the Tatra Mountains is preserved in the upper part of the Rovienky Valley (Fig. 1), and was therefore chosen for a case study. Field work reveals an interesting spatial relation between moraines of the same age, whose geometry indicates a simultaneous occurrence of three small marginal glaciers in the valley. Well-developed and pronounced moraines, with the inner depression reaching a depth of over a dozen meters, were formed in the cirque backwalls during the last recessional stage.

We combined detailed geomorphological mapping supported by the Schmidt hammer dating method with geochronological data from [10]Be surface exposure dating. In order to obtain an overall insight into the time and pattern of the final deglaciation of the Tatra Mountains the approach was applied during the preliminary

reconnaissance. Seven samples, five from boulders (Fig. 2) and two from ice-polished bedrock, were taken for dating with [10]Be from Rovienky Valley. The area is built of granitoides and is lithologically uniform, prone to provide reliable results for Schmidt hammer dating, which was conducted on each site intended for cosmogenic nuclide dating.

Fig. 2: *Boulder on the top of the Lateglacial moraine selected for dating (white arrow) in the forefront of the most prominent moraine.*

Our preliminary results suggest that the youngest and most prominent moraine system was formed during the Younger Dryas readvance. All calculated exposure dates yield ages between 13 ka and 11 ka. Ongoing work aims to obtain additional [10]Be exposure ages which in correlation with relative chronological data and morphostratygraphy will allow to determine the pattern and chronology of the massif deglaciation.

[1] J. Zasadni and P. Kłapyta, Geomorph. 253 (2016) 406

[1] *Geology, AGH University, Kraków, Poland*
[2] *Geography, Jagiellonian University, Kraków, Poland*
[3] *Earth Sciences, University of Pisa, Italy*

LATEGLACIAL LANDSCAPE DEVELOPMENT IN MEIENTAL

Geomorphic mapping and [10]Be dating of selected glacial landforms

M. Boxleitner[1] , M. Maisch[1], S. Ivy-Ochs, D. Brandova[1], M. Christl, M. Egli[1]

After the Last Glacial Maximum (LGM), several glacier readvances occurred characterized by distinct moraine complexes along mountain valleys and in cirques. Generally, the moraines of the multi-phased Egesen stadial are easily recognizable and can be found in many valleys [1]. For older stages, however, age constraints are rather rare. The aim of our study in Central Switzerland is to develop a refined relative morphostratigraphy and absolute chronology of the local glacial development and the landscape history of the Meiental (Canton of Uri) after the LGM using a multimethodological approach.

The selection of the sites was based on a geomorphic map [2]. Due to the presence of several moraines that are considered to have been deposited in the Oldest Dryas (Daun stadial or older), we have the unique possibility to derive a detailed sequence for the entire Lateglacial period. Following geomorphic mapping, more than 20 rock samples were collected from moraines for [10]Be analysis. First results show that the moraines can be attributed to glacier readvances of the Oldest and Younger Dryas (Figs. 1 & 2), i.e. the time span between 10-16 kyrs BP.

Fig. 2: *Lateglacial moraine (see triangle).*

[10]Be exposure dates were calculated using the NENA production rate [3] with an erosion rate of 1 mm/ky without snow correction. Together with additional [10]Be and ^{14}C samples from the Meiental, Göschenertal, Upper Engadine and published data, sequences will be completed. This will help to derive improved estimates of equilibrium lines of altitude (ELA) of glaciers from the LGM to the beginning of the Holocene.

[1]   Moran et al., Holocene (2015)
[2]   F. Renner, Phys. Geogr. 8 (1982)
[3]   Balco et al., Quat. Geochron. 4 (2009) 2

Colour	Local glacier extent	East-alpine equivalent
	1850	1850
	Chalchtal	Egesen
	Meien Dörfli	Daun
	Wassen	Clavadel?

Fig. 1: *Mapped moraines, local glacial morpho-stratigraphy and sample locations in the upper Meiental with selected [10]Be-ages.*

---

[1] *Geography, University of Zurich*

LATE HOLOCENE EVOLUTION OT THE TRIFTJEGLETSCHER

Combination of mapping, surface exposure and radiocarbon dating

O. Kronig, S. Ivy-Ochs, I. Hajdas, M. Christl, C. Schlüchter

To develop a better understanding of glacier fluctuation in the Alps during the Holocene, the forefields of the Triftje- and the Oberseegletscher east of Zermatt, in the Valais Alps, were investigated.

Fig. 1: *Overview of the study area (reproduced with authorisation of swisstopo (JA100120)), showing the glaciers, their forefields, and the sample locations of the exposure-dated boulders (orange dots) and the radiocarbon dated wood (green dot). For scale, the left lateral moraine of the Findelgletscher is approximately 2.5 km long.*

A multidisciplinary approach of detailed geological and geomorphological observations, ^{10}Be exposure and radiocarbon dating were applied. Based on the gained results, an early Holocene glacial stage at the end of the Younger Dryas cold period, when first parts of the study area became ice free, was documented by several exposure ages from glacially polished bedrock samples (VT4, 6, 12) and two perched boulders (VT5, 11). No landforms attributable to the mid-Holocene were observed or dated. Either no glacial landforms were formed due to the generally warm climate and/or the landforms were overrun or destroyed by younger and more extensive advances.

Fig. 2: *Close up of sample position of VT7 and VT8 on a small moraine that was partially buried by the big right lateral of the Oberseegletscher (reproduced with authorisation of swisstopo (JA100120)).*

During the early late Holocene, several exposure ages underpin the field observations that suggest significant glacier advances (VT2, 7, 8, 12) older than the Little Ice Age (LIA). The ages show that they occurred during the Göschenen II stadial [1,2].

During the LIA, glacier advances built most of the landforms that shape the scenery nowadays. For those advances we obtained one exposure age of the right lateral moraine of the Oberseegletscher (VT1) and a radiocarbon age of wood entrained in the glacial sediment.

The youngest landforms are small (1 m) moraine ridges which rim the two lakes (ober and Blâw See) that formed during readvances of the glaciers in the 1920s and 1980s.

[1]   H. Zoller, Verh. Nat.forsch. Ges. Basel 77 (1966) 97

[2]   S. Ivy-Ochs et al., Quat. Sci. Rev. 28 (2009) 2137

THE LATEGLACIAL IN RHAETIAN ALPS

Dating Early Holocene and Egesen in Ortles-Cevedale Group with ^{10}Be

C. Baroni[1], M.C. Salvatore[1], S. Casale[2], G. Guidobaldi[2], S. Ivy-Ochs, M. Christl, C. Wirsig, A. Carton[3], L. Carturan[4]

The Rhaetian Alps are a key site for Pleistocene paleoclimatic reconstructions providing several proxy data for investigating Lateglacial and Holocene glacial history. In particular, the Ortles-Cevedale Group, hosting the largest glacier covered massif of the Italian Alps, preserve several geomorphological evidences for characterizing the main steps of alpine glacial history [1].

The Pejo Valley is located in the southeastern sector of the Ortles-Cevedale Group and today hosts the La Mare and Careser glaciers [2, 3].

Several well-defined moraines allow us to identify different Lateglacial phases and to reconstruct the glacial history of this massif. Previous studies [3] identified the Holocene maximum extension of La Mare Glacier dated at about 1600 AD. In order to chronologically constrain the main steps of Lateglacial phases in the area, samples of erratic boulders (Fig. 1) from selected moraines were collected and dated with ^{10}Be.

Fig. 1: Erratic boulder.

We processed samples related to Lateglacial positions of La Mare Glacier (LAM13_1/6) and Careser Glacier (POVE 1,2,3) (Fig. 2).

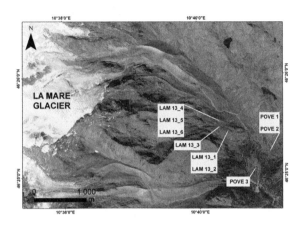

Fig. 2: Sample location map.

The ages obtained constrain Early Holocene and Lateglacial phases. One identified Lateglacial phase refers to the Egesen stadial correlated to the Younger Dryas.

These results, related to the geomorphological, geological and glacial settings of the area, give a more detailed knowledge of the La Mare glacier and Careser glacier history. New data are reinforced by already available ^{14}C ages for the same area [3] and are also in agreement with exposure ages in the nearby Rabbi Valley [4].

[1] G.N.G.F.G., Geogr. Fis. Dinam. Quat. 9 (1986) 137
[2] L. Carturan et al., The Cryosphere 7 (2013) 1819
[3] L. Carturan et al., Geogr. Ann. Ser. A-physi. Geogr. (2014) 287
[4] F. Favilli et al., Geomorphology 112 (2009) 48

[1] Earth Sciences, University of Pisa, Italy
[2] Regional Doctoral School in Earth Sciences, University of Pisa, Italy
[3] Geoscience, University of Padova, Italy
[4] Land, Environment, Agriculture and Forestry, University of Padova, Italy

REACHING AND ABANDONING THE FURTHEST ICE EXTENT

Overview of timing of the Last Glacial Maximum in the Alps

S. Ivy-Ochs, C. Wirsig, K. Hippe, M. Christl

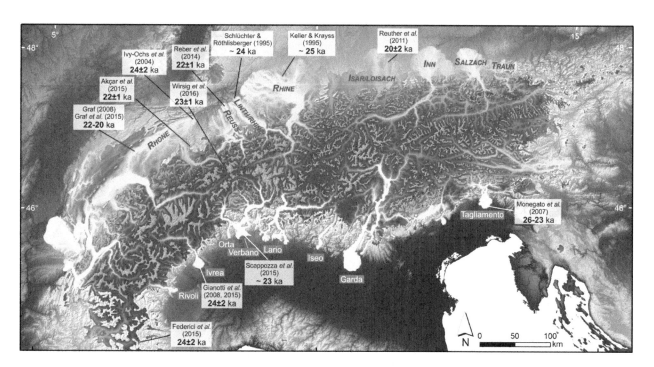

During the Last Glacial Maximum (LGM) in the European Alps (late Würm) local ice caps and extensive ice fields in the high Alps fed huge outlet glaciers that occupied the main valleys and extended onto the forelands as piedmont lobes (see figure above). Records from sites on both the northern and southern side of the Alps suggest advance of glaciers beyond the mountain front at around 30 ka (for complete references see [1]). Reaching of the maximum extent occurred by no later than 27-26 ka, as exemplified by dates from the Rhine glacier area [2]. Abandonment of the outermost moraines at sites north and south of the Alps was underway by about 24 ka.

In the high Alps, systems of transection glaciers with transfluences over many of the Alpine passes dominated, for example, at Grimsel Pass. ^{10}Be exposure ages of (23 ± 1) ka for glacially sculpted bedrock located just a few meters below the LGM trimline in the Haslital near Grimsel Pass suggest a pulse of ice surface lowering at about the same time as the foreland moraines were being abandoned [3]. Thereafter, glaciers oscillated at stillstand and minor readvance positions on the northern forelands and within the Italian amphitheatres for several thousand years forming the LGM stadial moraines. Final recession back within the mountain front took place by 19-18 ka.

Recalculation to a common basis of all published ^{10}Be exposure dates for boulders situated on LGM moraines suggests a strong degree of synchrony for the timing of onset of ice decay both north and south of the Alps.

[1] S. Ivy-Ochs, Cuadernos de investigación geográfica 41 (2015) 295

[2] O. Keller and E. Krayss, Vierteljahrschr. Naturforsch. Gesell. Zürich 150 (2005) 69

[3] C. Wirsig et al., J. Quat. Sci. 31 (2016) 46

THE LAST RETREAT OF THE REUSS GLACIER

Exposure ages indicate the timing of LGM downwasting

R. Reber[1] N. Akçar[1], S. Ivy-Ochs, D. Tikhomirov[2], R. Burkhalter[3], P.W. Kubik, C. Vockenhuber, C. Schlüchter[1]

The chronology of frontal and lateral moraines of the Reuss Glacier in the northern Alpine foreland was studied by surface exposure dating of erratic boulders with cosmogenic ^{10}Be and ^{36}Cl. We conclude that the exposure ages from the lateral Seeboden moraine on the Rigi indicate an ice surface at an elevation of about 1,000 m a.s.l. until (20.4 ± 1.0) ka [1]. The onset of the Last Glacial Maximum (LGM) could be established by comparing moraines in terminal and retreat positions (Fig. 1).

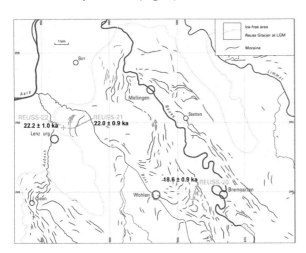

Fig. 1: *Glacial morphological map of the frontal part of the LGM Reuss Glacier [1].*

Our results indicate that the Reuss Glacier started to retreat at (22.2 ± 1.0) ka (Reuss-22) which seems to be synchronous with the retreat of the Valais and the Aare Lobes (Fig. 2) further to the west. The Reuss lobe retreated for 12 km to Wohlen from the frontal position (in 2-3 ka, Reuss-20 = (18.6 ± 0.9) ka; Figs. 1 and 2). After (18.6 ± 0.9) ka, rapid ice decay or even collapse of the foreland glaciers may have occurred. Lateglacial glacier advances happened only in the high-alpine valleys, e.g. during the Gschnitz stadial at (17.1 ± 1.6) ka [2]. The equilibrium line altitude (ELA) was at least 500 m higher during

the Gschnitz stadial than it was during the LGM [2]. Furthermore, our data is pointing towards a simultaneous deglaciation from frontal (Lenzburg) and lateral (Seeboden) position, which requires an active down-melting of the ice [1].

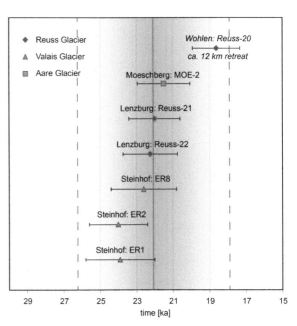

Fig. 2: *Compilation of exposure ages from the northern Alpine foreland showing data from the Reuss Glacier [1], the Valais Glacier [2] and the Aare Glacier [3].*

[1] R. Reber et al., Swiss J. Geosci. 107, 2-3 (2014) 293

[2] S. Ivy-Ochs et al., J. Quat. Sci. 23, 6-7 (2008) 559

[3] N. Akçar et al., Swiss J. Geosci. 104 (2011) 445

[1] *Geology, University of Bern*
[2] *Physics and Astronomy, Aarhus University*
[3] *swisstopo, Wabern*

LATE PLEISTOCENE GLACIERS IN NORTHERN APENNINES

In situ cosmogenic nuclides as a tool for constraining glacial history

C. Baroni[1], G. Guidobaldi[2], M.C. Salvatore[1], M. Christl, C. Wirsig, S. Ivy-Ochs

Northern Apennines (Italy) occupy a strategic position between the Alpine region and the Mediterranean basin, strongly influenced by the nearby Ligurian Sea. During the Last Glacial Maximum (LGM) several valley glaciers occupied and shaped the head of the main Apenninic valleys. Geomorphological evidences also testify to distinct glacial phases after LGM.

[10]Be exposure ages are obtained for the first time in the Northern Apennines. Samples from erratic boulders were collected in selected key sites to outline the glacial history of the mountain range during the Late Pleistocene.

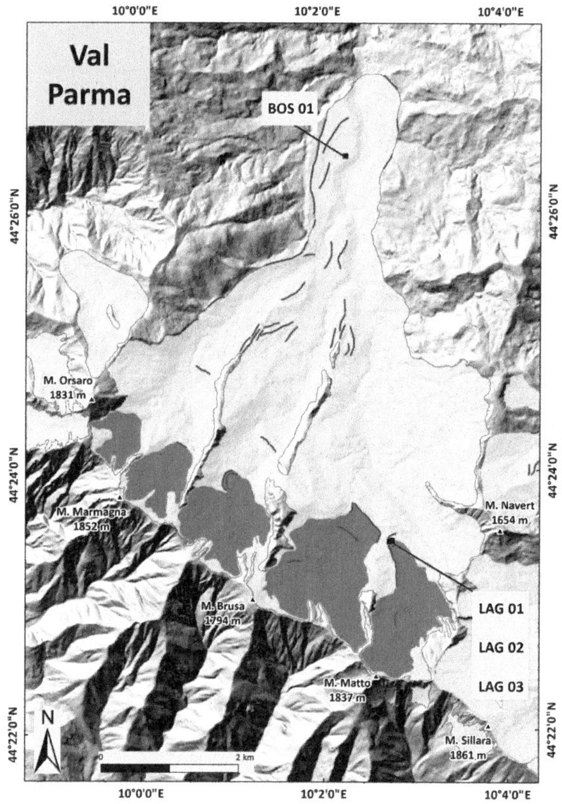

Fig. 1: *Val Parma Glacier during LGM (in blue) and the last Lateglacial phase (in purple).*

[10]Be dated samples range from the LGM to the last Lateglacial phase, tentatively related to the Younger Dryas. During the LGM, Val Parma (Fig. 1) was occupied by a 23 km[2] complex valley glacier with a well-defined tongue. After the LGM, the glacier fragmented in distinct glacial bodies and progressively retreated until reaching the cirque phase during the Younger Dryas.

About 20 km southeast from Val Parma the valley glacier that developed from Monte La Nuda (Fig. 2) provides evidences of at least three Lateglacial phases.

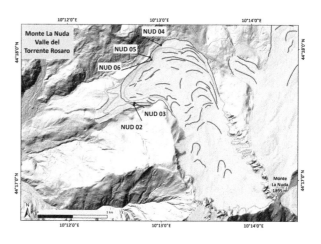

Fig. 2: *Monte La Nuda Glacier during the LGM and early Lateglacial phases (blue and orange).*

Results obtained fall within the scope to reconstruct timing and entity of the last glaciation along the Northern Apennines during the Late Pleistocene. They contribute to outline the paleoclimatic dynamics that characterized the investigated area in the framework of the Western Mediterranean.

[1] *Earth Sciences, University of Pisa, Italy*
[2] *Regional Doctoral School in Earth Sciences, University of Pisa, Italy*

SYNCHRONOUS LAST GLACIAL MAXIMUM

LGM equilibrium line altitude depression across Turkey

N. Akçar[1], V. Yavuz[2], S. Yeşilyurt[3], S. Ivy-Ochs, R. Reber[1], C. Bayrakdar[4], P.W. Kubik, C. Zahno[5], F. Schlunegger[1], C. Schlüchter[1]

Uludağ is a solitary mountain in NW Turkey where glacial deposits have been documented in the Kovuk Valley and where the glacial history has been reconstructed based on thirty-one cosmogenic [10]Be exposure ages from glacially transported boulders and bedrock. The Kovuk Glacier began advancing before (26.5 ± 1.6) ka. It reached its maximum extent at (20.3 ± 1.3) ka, followed by a readvance at (19.3 ± 1.2) ka, both during the global Last Glacial Maximum (LGM) within Marine Isotope Stage-2 [1]. Based on the geomorphologic ice margin reconstruction and using the accumulation-ablation area ratio (AAR) approach, the LGM equilibrium line altitude (ELA) of the Kovuk LGM-maximum glacier was ca. 2000 m above sea level for an estimated AAR of 0.67.

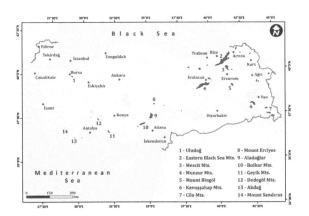

Fig. 1: *Maximum extent of glaciers in the Anatolian mountains during the Quaternary [1].*

Based on the lower bounds of the modern ELA estimates, we tentatively estimated the ELA depressions for the investigated palaeoglaciers during the LGM. The ELA depression was approximately 800 – 1000 m on Uludağ, ca. 600 – 800 m in the eastern Black Sea Mts., ca. 1300 m on Mt. Erciyes, ca. 1100 m in the Dedegöl Mts., ca 900 m on Akdağ, and ca. 1000 m on Mt. Sandıras [1].

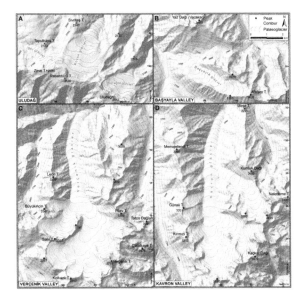

Fig. 2: *Reconstructed LGM glaciers in Uludağ and in the eastern Black Sea Mountains [1].*

The LGM in the Anatolian mountains can be characterized by glaciers that responded to MIS-2 cooling and reached their maximum position at (20.9 ± 1.5) ka. The lowering of the ELA of these glaciers was on the order of 1000 m compared to the modern ELA estimates. The maximum extent of Anatolian glaciers appears to be synchronous with the LGM advances in other mountains of the Mediterranean basin [1].

[1]   N. Akçar et al., Geol. Soc. London Spec. Publ. 433 (2015)

[1] *Geology, University of Bern*
[2] *Geological Engineering, Istanbul Technical University, Turkey*
[3] *Geography, Ankara University, Turkey*
[4] *Geography, Istanbul University, Turkey*
[5] *CH-6403 Küssnacht*

TIMING OF FLUVIAL TERRACE FORMATION IN EUROPE

Age constraints by cosmogenic nuclide dating

M. Schaller[1], T.A. Ehlers[1], T. Stor[2,3], J. Torrent[4], L. Lobato[5], M. Christl, Ch. Vockenhuber

Age constraints of late Cenozoic fluvial deposits are important to address questions in geomorphology such as incision rates of rivers. Fluvial sequences are often dated with relative age constraints such as magneto- or biostratigraphy. Absolute age constraints of fluvial deposits are not always possible as datable material is missing (e.g. no organic matter for carbon dating) or techniques do not cover the age ranges of the fluvial deposits (e.g., carbon dating in early Pleistocene terraces). Different dating techniques based on in situ-produced cosmogenic nuclides allow the age determination of fluvial deposits back to 5 Myr.

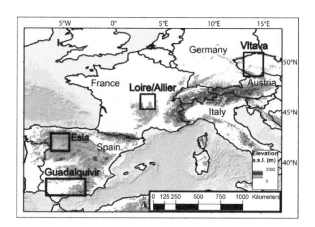

Fig. 1: *Location of four terrace sequences in Europe analyzed for isochron burial dating and depth profile dating with cosmogenic nuclides.*

In this study, cosmogenic depth profile dating and isochron burial dating were applied to four different river systems in Europe spanning 12° of latitude. Isochron burial age constraints of four selected terraces from the Vltava river (Czech Republic) range between (961 ± 225) to (2086 ± 507) kyr. An isochron burial age derived for the Allier river (Central France) is (2000 ± 223) kyr. Two terrace levels from the Esla river (NW Spain) were dated to (155 +62/-6) kyr and (586 +128/-198) kyr with

depth profile dating. The latter age agrees with an isochron burial age of (524 ± 195) kyr. The successfully dated terrace level from the Guadalquivir river (SW Spain) resulted in an isochron burial age of (1787 ± 188) kyr.

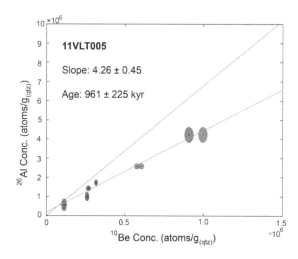

Fig. 2: *Isochron burial dating of sample 11VLT005 from the terrace sequence of the Vltava river, Czech Republic. The uncorrected values (blue ellipses) are linearized (grey ellipses). The slope through the linearized sample indicates an age of (961 ± 225) kyr.*

Results indicate that the cosmogenic nuclide-based ages are generally older than the ages derived from previous relative age constraints. These results highlight a potential uncertainty in relative age constraints used to understand climate drivers for terrace formation within Europe and elsewhere.

[1] *Geosciences, University of Tuebingen, Germany*
[2] *Geology, Charles University Prague, Czech Republic*
[3] *Czech Geological Service, Czech Republic*
[4] *Agronomy, University of Córdoba, Spain*
[5] *Geography and Geology, University of León, Spain*

PAST LANDSCAPE CHANGE IN THE HOHEN TAUERN (A)

Glacial and periglacial activity before the Bølling/Allerød interstadial

J.M. Reitner[1], S. Ivy-Ochs, O. Kronig, A.U. Reuther[2], M. Christl

The Gschnitz stadial was the first ice re-advance after the LGM during the Alpine Lateglacial when glaciers had ice-free conditions in the forefield. Based on previous exposure dating of a terminal moraine [1] it is known that this major glacier advance occurred around 16 ka during Heinrich event 1 which resulted in a major cooling phase. However, until now no other location tentatively tied to the Gschnitz stadial has been directly dated so far in the Alps. All other sites were correlated based only on morphological features and/or on typical changes of the equilibrium line altitude.

Fig. 1: *Situation in the Malta valley with reconstructed paleo-glacier.*

In the Malta valley (in Carinthia) there is a textbook example of terminal moraines confining a small tongue basin. These moraines are linked to fluvial terraces indicating strong aggradation under free drainage (ice-free) conditions in the forefield of the paleo-glacier terminus. Due to the similarities with the Gschnitz type - locality this situation was perfect to test whether the assumption of simultaneous formation is true. Three orthogneiss boulders on the lateral moraines were sampled. First results show that the glacier advance of Malta occurred during the Gschnitz stadial.

Paleo-permafrost investigations and the reconstruction of permafrost degradation in the Alps during the Lateglacial rely mostly on geomorphic-geological features such as relict rock glaciers. In general the elevation of a rock glacier's terminus is regarded as providing an indicator of the lower limit of discontinuous permafrost for the period when it was active. In the Malta valley findings of a series of very low reaching rock glacier deposits, i.e. up to 1300 m below modern permafrost limits in these areas [1], make formation of such features even prior to the Younger Dryas plausible. The recent ^{10}Be dating results of two boulders indeed show that these stabilized around the transition from Oldest Dryas stadial to the Bølling/Allerød interstadial. Further work dating the rock glaciers is being done in the framework of an SNF-funded project.

Fig. 2: *Surface of one of the relict rock glaciers.*

[1] S. Ivy-Ochs, et al., J. Quat. Sci. 21 (2006) 115

[2] J.M. Reitner, Jb. Geol. Bund. Wien 147 (2007) 672

[1] *Geological Survey of Austria, Vienna, Austria*
[2] *University of Colorado, Boulder, USA*

PAST ROCK GLACIER ACTIVITY IN THE NORTHERN ALPS

^{36}Cl dating of relict rock glaciers, Northern Calcareous Austrian Alps

A.P. Moran[1], S. Ivy Ochs, C. Vockenhuber, H. Kerschner[1]

Rock glaciers are geomorphological features in mountainous environments showing evidence of discontinuous permafrost activity [1]. After the meltout of permafrost under warming conditions, relict rock glaciers can still be identified as bouldery tongue or lobate shaped landforms. They represent valuable climate proxies as they provide evidence of past permafrost distribution [2], which is itself a function of mean annual air temperature [3].

In the Northern Calcareous Alps of Austria relict rock glaciers are currently being investigated. They are located in the Karwendel Mountains on the north side of the "Nordkette" near Innsbruck (Tyrol) in an altitudinal belt reaching from 2100-1950 m a.s.l. Their size infers a time of activity of several hundred years. According to their altitude in association with nearby moraine systems, they can be expected to have formed during the Younger Dryas cold period (~12.8-11.7 ka) or the early Holocene [4] and stabilized during the onset of warm interglacial conditions.

Fig.1: *Relict rock glacier in the Karwendel Mountains, Austria.*

As Wetterstein Limestone dominates in the research area, we use surface exposure dating

with ^{36}Cl of boulders on the rock glaciers to determine the age of their final stabilization. It will allow for the first time an association of the period of activity of these landforms in the Northern Alps to a specific climatic period. They also provide first chronological information for the glacial and periglacial deposits in the Northern Alps.

By comparing the altitude of the lower limit of past permafrost with the lower limit of current permafrost, a shift in mean annual air temperature between the time of past activity and present conditions can be estimated.

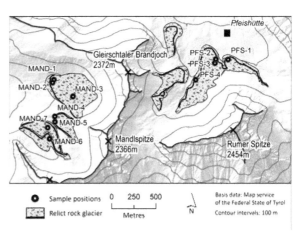

Fig. 2: *Relict rock glaciers and their sampling sites in the Karwendel Mountains, Austria.*

[1] W. Haeberli, Mitt. VAWHG ETHZ 77 (1985)
[2] H. Kerschner, Arct. Alp. Res. 10 (1978) 635
[3] D. Barsch: Rock Glaciers. Springer, Berlin (1996)
[4] H. Kerschner, Innsbr. Geogr. Stud. 20 (1993) 47

[1] *Geography, University of Innsbruck*

ROCK AVALANCHES IN THE MONT BLANC MASSIF (ITALY)

Chronology of repeated rock avalanches onto the Brenva Glacier

P. Deline[1], N. Akçar[2], S. Ivy-Ochs, P.W. Kubik

Infrequent rock avalanches (volume greater than $10^6 \, m^3$) are long-runout processes that may threaten populated mountain valleys. Rock avalanches also have strong implications for relief generation and destruction over time. Here we propose a chronology for seven of the rock-ice avalanches that affected a steep glacier basin on the southeast side of the Mont Blanc during the late Holocene (Fig. 1). A geomorphological study of the runout deposits on the valley floor and the opposite side was combined with the analysis of historical sources and the use of absolute and relative dating methods, especially surface exposure dating with cosmogenic ^{10}Be of 18 granite boulders from two deposits [1].

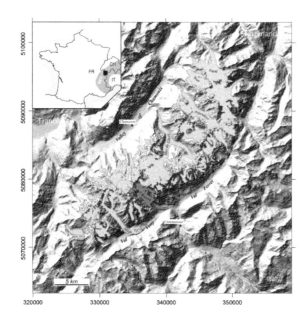

Fig. 1: *Location of the Mont Blanc massif and shaded-relief map showing the present-day glacier extent in the massif [1].*

These rock-ice avalanches are dated AD 1997 and 1920, with a rock volume in the range 2.4-3.6 and 2 x $10^6 \, m^3$, respectively; AD 1767, with a slightly shorter runout; AD 1000-1200, with a

longer runout; ca. AD 500, the runout of which is uncertain; ca. 2500 years BP, the determination of which is indirect; and ca. 3500 years BP (Fig. 2), with the longest runout. [1].

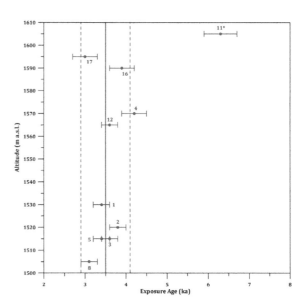

Fig. 2: *Exposure ages of the boulders from the ca. 3500 years BP Brenva rock avalanche deposit [1].*

There is no distinct relationship between climatic periods and the occurrence of these rock avalanches. Even for the two best documented ones. Modelling suggests that the 1997 scar was characterized by permafrost close to 0°C, whereas, the 1920 scar was located in cold permafrost [1].

[1] P. Deline et al., Quat. Sci. Rev. 126 (2015) 186

[1] *EDYTEM Lab, Savoie University, France*
[2] *Geology, University of Bern*

THE MID-HOLOCENE INCREASED NEOTECTONIC ACTIVITY

The Sennwald landslide using surface exposure dating with ^{36}Cl

S. Aksay[1], S. Ivy-Ochs, K. Hippe, L. Grämiger[1], C. Vockenhuber

The Säntis thrust is a fold-and-thrust structure in eastern Switzerland consisting of numerous tectonic discontinuities that make the bedrock vulnerable to rock failure. The Sennwald landslide is one of these events that occurred due to the failure of Lower Cretaceous Helvetic limestones in the Rhine valley. In this study, the Sennwald landslide is displayed with the surface exposure age in relation with geological and tectonic setting, earthquake frequency, and regional scale climate/weather influence.

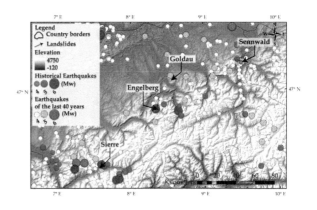

Fig. 2: *The earthquake map of Swiss Alps [2]*

The Sennwald landslide is likely to have been triggered by earthquake activity. The exposure ages imply that the rock failure occurred during the middle Holocene, a period of increased neotectonic activity in Eastern Alps [3] (Fig. 2). The last 40 year's earthquake activity and historical earthquakes (M ~ 4.0-6.0) also show that this region is tectonically still active (Fig. 2). This time period also coincides with notably wet climate, which has been suggested as an important trigger for landslides around this age across the Alps [4].

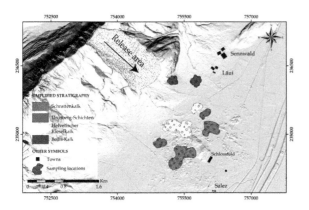

Fig. 1: *Sampling locations and the distribution of deposits in the release area.*

During the landslide the original bedrock stratigraphy was roughly preserved as geologically the top layer in the bedrock package travelled the farthest and the bottom layer came to rest closest to the release bedrock wall. Total Cl and ^{36}Cl were analyzed at the ETH AMS facility with isotope dilution methods defined in the literature [1]. Surface exposure ages of landslide deposits in the accumulation area are determined from twelve boulders.

The distribution of limestone boulders in the accumulation area (Fig. 1), the numerical run-out modelling and in particular the exposure ages support the hypothesis that the landslide was a single catastrophic event.

[1] S. Ivy-Ochs et al., Nucl. Instrum. Methods Phys. Res. B223 (2004) 623

[2] D. Fäh et al., SED ETH Zurich, Earthquake Catalogue of Switzerland (2011)

[3] Prager et al., Nat. Hazards Earth Syst. Sci. 8 (2008) 377

[4] S. Zerathe et al., Quat. Sci. Rev. 90 (2014) 106

[1] *Geological Institute, ETH Zürich*

DATING THE SENTINEL ROCK AVALANCHE OF ZION, UTAH

Surface exposure dating of landslide deposits with ^{10}Be

J. Castleton[1], J.R. Moore[1], M. Christl, S. Ivy-Ochs

Blocking the mouth of Zion Canyon, Utah, over a distance of >3 km, the prehistoric Sentinel rock avalanche has had long-lasting impact on the iconic scenery of Zion National Park, once damming a large lake that filled the rocky canyon floor with sediment [1]. The massive failure also represents an extreme-magnitude hazard for Zion's nearly 3 million annual visitors.

Fig. 1: *A) View over rock avalanche deposits in Zion Canyon. B) Section of Sentinel rock avalanche debris exposed by river incision. C) Lacustrine clay beds exposed near Zion Lodge. D) Partial view of the Sentinel source area.*

We combine new mapping of rock avalanche and related lacustrine deposits to reconstruct topography before and after the landslide, comment on failure kinematics, and determine refined volume estimates. Cosmogenic nuclide surface exposure dating of deposited rock avalanche boulders allows us to provide the first direct age of the slide, determine rates of erosion, and explore potential triggering mechanisms. Boulders from across the slide surface were deposited simultaneously, yielding similar exposure ages and indicating a single massive and catastrophic rock slope failure [2].

Fig. 2: *Sampling a boulder of Navajo Sandstone for surface exposure dating.*

Geological evidence shows that following the slide, Zion Canyon contained a lake for several centuries until eventually filling with sediment, creating the modern-day flat valley floor. Long-lasting geomorphic and ecological effects attest to the diverse impacts of large rock avalanches in steep desert landscapes.

[1]   R.K. Grater, J. Geol. 43 (1945) 116
[2]   J. Castleton et al., GSA Today (2016) in press

[1] *Geology and Geophysics, University of Utah, USA*

DATING HOLOCENE BASALTS AND HUMAN FOOTPRINTS

Surface exposure dating of mafic lava flows using ^{3}He and ^{10}Be

C. Heineke[1], R. Hetzel, [1]S. Niedermann[2], C. Akal[3], M. Christl

The Kula volcanic field (western Turkey) comprises about 80 cinder cones and associated basaltic lava flows of Quaternary age. Human footprints in ash deposits document that the early inhabitants of Anatolia were affected by the volcanic eruptions, but the age of the footprints and the most recent volcanic activity remains poorly constrained [1]. To resolve the timing of the latest volcanic activity we determined the age of the youngest lava flows and cinder cones by applying ^{3}He and ^{10}Be exposure dating to olivine phenocrysts in basalt and quartz-bearing xenoliths [2].

Fig. 1: *Holocene cinder cone and lava flows of the Kula volcanic field (western Turkey).*

Based on geomorphological criteria and K-Ar dating three eruption phases can be distinguished in the Kula volcanic field [3, 4]. In order to constrain the age of the latest phase of volcanism, we collected samples from basaltic lava flows and cinder cones assigned to the most recent group of volcanic deposits (Fig. 1) [2, 3]. In addition, three metamorphic xenoliths were collected for ^{10}Be dating. Two xenoliths were discovered on a lava flow and a small cinder cone, respectively. The third one was obtained from the top of a large cinder cone, whose youngest ash deposits contains the well preserved human footprints (Fig. 2).

Both, ^{3}He and ^{10}Be exposure dating yielded consistent results for the Holocene volcanic rocks of the Kula volcanic field. Exposure ages from the lava flows and the small cinder cone cluster between ~1.4 and ~3.1 ka, indicating a pronounced volcanic activity that began in already historic times [2]. For the xenolith of the large cinder cone a ^{10}Be age of ~11 ka provides the first robust age constraint for the famous human footprints that are preserved in the associated volcanic ash deposits [2].

Fig. 2: *Human footprint in volcanic ash.*

Our results demonstrate that cosmogenic nuclides can be applied to date very young volcanic rocks, even in regions of low elevation where nuclide production rates are low. Hence, exposure dating provides a powerful alternative to K-Ar or ^{40}Ar/^{39}Ar dating of young volcanic rocks [2].

[1] W. Barnaby, Nature 254 (1978) 553

[2] C. Heineke et al., Quat. Geochron. (submitted)

[3] J.-M. Richardson-Bunbury, Geol. Mag. 133 (1996) 275

[4] E. Şen et al., Bull. Earth. Sci. Appl. Res. Cent. Hacettepe Univ. 35 (2014) 219

[1] *Geology and Paleontology, University of Münster, Germany*

[2] *Anorganic and Isotope Geochemistry, GFZ Potsdam, Germany*

[3] *Geological Engineering, University of Izmir, Turkey*

ISOCHRON-BURIAL DATING: TERRACE CHRONOLOGIES

Quaternary uplift rates of the Central Anatolian Plateau, Turkey

A. Çiner[1], U. Doğan[2], C. Yıldırım[1], N. Akçar[3], S. Ivy-Ochs, V. Alfimov, P.W. Kubik, C. Schlüchter[3]

The Central Anatolian Plateau (CAP) in Turkey is a relatively small plateau (300 x 400 km) with moderate average elevations of ca. 1 km and is situated between the Pontide and Tauride orogenic mountain belts. Kızılırmak, which is the longest river (1355 km) within the borders of Turkey, flows within the CAP and slowly incises into lacustrine and volcaniclastic units before finally reaching the Black Sea. We dated the Cappadocia section of the Kızılırmak terraces in the CAP by using burial and isochron-burial dating methods with cosmogenic ^{10}Be and ^{26}Al (Fig. 1). Absolute dating of the terraces provides insight into long-term incision rates, uplift and climatic changes [1].

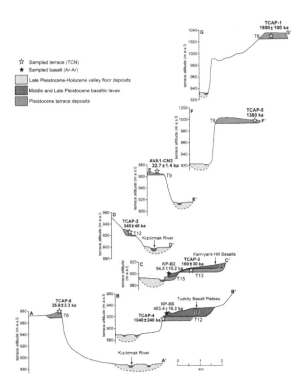

Fig. 1: *Reconstructed chronologies of the Kızılırmak River [1].*

Terraces at 13, 20, 75 and 100 m above the current river indicate an average incision rate of

0.051 ± 0.01 mm/a since 1.9 Ma. Using the base of a basalt flow above the modern course of the Kızılırmak, we also calculated 0.05-0.06 mm/a mean incision and hence rock uplift rate for the last 2 Ma (Fig. 2) [1].

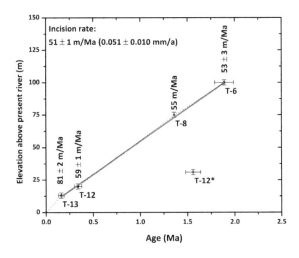

Fig. 2: *Reconstructed average incision rate [1].*

Although this rate might be underestimated due to normal faulting along the valley sides, it perfectly matches our results obtained from the Kızılırmak terraces. Even though it is up to 5-10 times slower, Quaternary uplift of the CAP is closely related to the uplift of the northern and southern plateau margins, respectively [1].

[1]   A. Çiner et al., Quat. Sci. Rev. 107 (2015) 81

[1] *Eurasian Earth Sciences, Istanbul Technical University, Turkey*
[2] *Geography, Istanbul University, Turkey*
[3] *Geology, University of Bern*

LITHOLOGY CONDITIONS SEDIMENT FLUX

In-situ [10]Be reveals variations in lithology-controlled sediment discharge

F. Cruz Nunes[1], R. Delunel[1], F. Schlunegger[1], N. Akçar[1], P.W. Kubik

We explore the influence of lithology on sediment discharge in the ca. 400 km^2 Glogn basin (Fig. 1), where the underlying bedrock dip is parallel to the topographic slope (dip slope situation) on the NW valley flank, while a non-dip slope situation is found on the opposite SE valley side. Accordingly, multiple landslides are perched on the dip slope side, while the opposite valley side hosts deeply dissected bedrock channels and threshold hillslopes.

This study [1] presents a [10]Be-based sediment budget to explore how the structural architecture of this region has conditioned the erosion pattern in the region. Our [10]Be based sediment budget (Fig. 2) suggests that ca. 60% of material has been derived through landsliding on the dip slope side, while the remaining 40% of sediment has been supplied from the opposite non-dip slope flank and upstream tributary basins [1]. This suggests that the tilt orientation of the bedrock exerts an important control on the erosional budget of a basin, mainly as the isoclinal tilt of the bedrock promotes landsliding on the dip slope facing valley side.

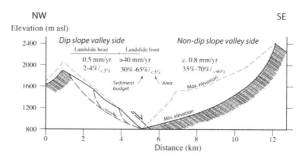

Fig. 2: *Geological architecture and Setting of the Glogn basin [1]. The [10]Be-based denudation rates and relative contributions to the basin's entire sediment budget along with the relative areal extents are shown.*

[1] F. Cruz Nunes et al., Terra Nova 27 (2015) 370

[1] *Geology, University of Bern*

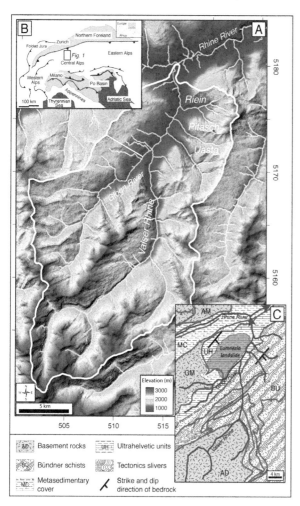

Fig. 1: *Glogn drainage basin, and lithotectonic architecture along with the bedding orientation of the bedrock [1].*

DENUDATION RATES OVER SPACE AND TIME IN EUROPE

Catchment-wide rates constrained by cosmogenic nuclides

M. Schaller[1], T.A. Ehlers[1], T. Stor[2,3], J. Torrent[4], L. Lobato[5], M. Christl, C. Vockenhuber

Constraints of denudation rates are important to study the interaction of climate, vegetation and tectonics with denudation. In situ-produced cosmogenic nuclide concentrations from river borne quartz grains allow the determination of catchment-wide denudation rates over space. The information of catchment-wide denudation rates is also stored in fluvial sedimentary records (e.g., the Vltava River in the Czech Republic; Fig. 1). These Paleo-denudation rates reflect possible changes in denudation over time.

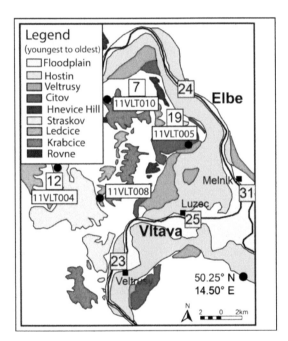

Fig. 1: River and terrace sediments of the Vltava (Czech Republic) analyzed for cosmogenic nuclides to determine denudation (yellow squares, in mm/kyr) and paleo-denudation rates (white squares, in mm/kyr).

In this study, in situ-produced cosmogenic nuclides from four river systems in Europe have been analyzed. Catchment-wide denudation rates derived from active river load range between 16 and 51 mm/kyr (Fig. 2). The scatter

in denudation rates is relatively large within a catchment area because of the different sizes and geomorphologic characteristics of analyzed catchments (e.g. Allier). Paleo-denudation rates are of the same order of magnitude, but generally slightly lower.

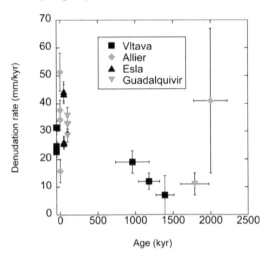

Fig. 2: Cosmogenic nuclide-derived denudation rates plotted against age from terrace sequences in Europe.

Results indicate that small catchments show more variability in denudation rates than large catchments. Therefore, for comparison of denudation rates over time, values from larger catchments should be compared (e.g. Vltava). However, reported cosmogenic nuclide-derived denudation rates from large catchments do not only smooth the denudation rates for geomorphic variability, but also for climate-induced changes.

[1] Geosciences, University of Tuebingen, Germany
[2] Geology, Charles Univ. Prague, Czech Republic
[3] Czech Geological Service, Czech Republic
[4] Agronomy, University of Córdoba, Spain
[5] Geography and Geology, University of León, Spain

DENUDATION RATES FROM ^{10}BE(METEORIC)/^{9}BE RATIOS

Testing the new proxy in small watersheds with variable lithology

N. Dannhaus[1], H. Wittmann[1], F. von Blanckenburg[1], P. Kram[2], M. Christl

We test the ^{10}Be(meteoric)/^{9}Be ratio as a proxy for denudation rates [1] on the scale of small creeks in the Slavkov Forest, Czech Republic. Meteoric cosmogenic ^{10}Be and its stable counterpart ^{9}Be mix to a characteristic ratio in the Critical Zone that is dependent on the depositional flux of meteoric ^{10}Be, the denudation rate D, the fraction of ^{9}Be released from primary minerals during weathering, and the ^{9}Be concentration of the parent bedrock [1]. This ratio can be determined on the reactive phase (adsorbed onto or precipitated in secondary minerals) of sediment which can be accessed using a sequential extraction method described in [2]. Unlike for *in situ*-produced ^{10}Be, determining denudation rates with the ^{10}Be(meteoric)/^{9}Be ratio does not depend on the presence of quartz.

We applied this new method that has recently been successfully tested on the large Amazon Basin scale [3] to three small catchments (< 1 km^2) in the Slavkov Forest. Each catchment is underlain by a different lithology, namely felsic granite (Lysina), mafic rocks like amphibolite (Na Zeleném), and serpentinite-rich ultramafic rocks (Pluhův Bor). These diverse lithologies are ideal to investigate the potential of the ^{10}Be/^{9}Be system under various geochemical conditions (e.g. variable pH, stream water chemistry).

We measured reactive ^{10}Be/^{9}Be ratios on fine-grained (< 63 µm) bedload sediment to calculate catchment-wide denudation rates (Fig. 1). Denudation rates for the felsic Lysina and for the amphibolite-dominated mafic Na Zeleném catchment are similar at 114 and 121 t km^{-2} y^{-1}, respectively. For the ultramafic serpentinite Pluhův Bor catchment the denudation rate is considerably lower at 55 t km^{-2} y^{-1}. The relative lower denudation rate is expected given the low erodibility of serpentinite.

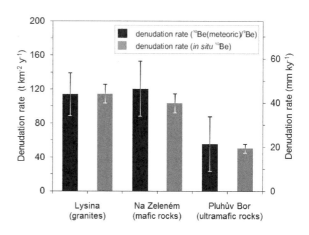

Fig. 1: *Denudation rates calculated from ^{10}Be(meteoric)/^{9}Be ratios (shown in black), and in situ-produced ^{10}Be (in gray) in t km^{-2} y^{-1} and mm ky^{-1}. The large uncertainties result from high ^{9}Be variability in the parent bedrock material.*

Sufficient amounts of quartz present in lenses and veins in the (ultra-)mafic catchments allowed comparison of these ^{10}Be/^{9}Be derived rates with *in situ* ^{10}Be denudation rates. Both methods agree within uncertainty (Fig. 1) and these new rates are further in the range of *in situ* ^{10}Be denudation rates for river catchments in middle Europe [4]. These promising results indicate that mixing of both isotopes has been accomplished at this small scale. A major advantage of this method is that it can be applied to any lithology, provided that the bedrock ^{9}Be concentration is known.

[1] von Blanckenburg et al., EPSL 295 (2012) 351

[2] Wittmann et al., Chem. Geol. 126 (2012) 318

[3] Wittmann et al., JGR 120 (2015)

[4] Schaller et al., EPSL 188 (2001) 441

[1] *Helmholtz Centre, GFZ Potsdam, Germany*
[2] *Czech Geological Survey, Prague, Czech Republic*

CONSTRAINING EROSION RATES IN SEMI-ARID REGIONS

What is the limiting factor for soil chemical weathering and erosion?

V. Vanacker[1], J. Schoonejans[1], S. Opfergelt[1], Y. Ameijeiras-Mariño[1], M. Christl

Arid and semi-arid environments occupy around 37% of the land surface. The irregular rainfall regime with prolonged dry periods supports only sparse and patchy native vegetation cover. Sustained development of rain-fed agriculture is limited by soil and water resources [1].

This study focuses on the relationship between soil production, physical erosion, and chemical weathering. Study sites are located in the Southern Betic Cordillera (SE Spain), and selected across a spatial gradient in climatic and topographic conditions. The strong contrasts between the gently sloping hillsides of the Sierra de las Estancias, and the steep, highly dissected landscape of the Sierra Cabrera reflects the tectonic history of the Betic ranges.

Fig. 1: *Sampling of soil profiles in the Sierra Estancias (southeast Spain).*

Four catchments were selected to cover the range of denudation rates that were established for the Betic Cordillera [2]. In each catchment, two to three regolith profiles were sampled on exposed ridgetops to avoid the complexities of

soil forming processes associated with lateral transport of chemical fluids and soil particles along slope (Fig. 1). Total elemental composition of soil and rock samples was determined by ICP – AES (Thermo iCAP 6000 Series), and soil and sediment samples were processed for in-situ cosmogenic ^{10}Be analyses.

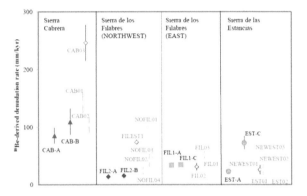

Fig. 2: ^{10}Be-derived denudation rates of soils (colored symbols) and sediments (black symbols) in the Betic Cordillera.

In the Southern Betic Cordillera, soil denudation rates are low, and range between 14 and 109 mm/kyr. Soil denudation rates are generally less than or equal to catchment-wide denudation rates measured at the outlet of small basins. Chemical weathering losses account for ~5 to 30 % of the total mass lost by denudation. Soil weathering increases (nonlinearly) with soil thickness and decreases with increasing surface denudation rates, consistent with kinetically limited weathering.

[1] V. Vanacker et al., Landscape Ecol. 29 (2014) 293

[2] N. Bellin et al., Earth Planet. Sci. Lett. 390 (2014) 19

[1] *Earth & Life Institute, University of Louvain, Belgium*

EROSION OF THE CENTRAL BOLIVIAN ANDES

Tectonic uplift and lithology controlling ^{10}Be-^{26}Al denudation rates

K. Hippe, F. Kober[1], G. Zeilinger[2], O. Marc[2], T. Lendzioch[2], R. Grischott[1], M. Christl, P.W. Kubik, R. Zola[3]

The topographic signature of a mountain belt depends on the interplay of tectonic, climatic and erosional processes. We investigate these processes in the Rio Grande catchment which crosses orthogonally the eastern Andes orogen from the Eastern Cordillera into the Subandean Zone (Fig. 1), exhibiting a catchment relief of up to 5000 m.

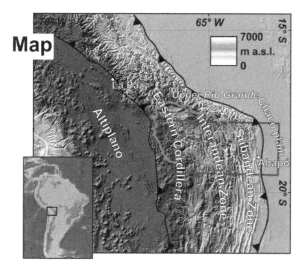

Fig. 1: *Location of the upper Rio Grande catchment on the eastern flank of the central Andes of Bolivia.*

Our dataset of 57 cosmogenic ^{10}Be and ^{26}Al catchment wide denudation rates from the Rio Grande catchment reveals up to one order of magnitude higher denudation rates in the Subandean Zone compared to the upstream physiographic regions [1]. Based on cumulative rock uplift investigations and due to the absence of a pronounced climate gradient, we infer that increased tectonic activity in the Subandean belt causes the higher denudation rates (Fig. 2). Despite the enhanced tectonic activity in the Subandes, local relief, mean and modal slopes and channel steepness indices are largely similar compared to the Eastern Cordillera and the intervening Interandean Zone. However, higher

denudation rates are also associated with lower rock-strength lithologies of the Subandean sedimentary units, showing that lithology and rock strength can control high denudation at low slopes.

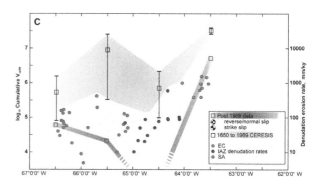

Fig. 2: *Calculated seismically uplifted volume (with error envelopes) showing higher uplift in the Subandes (SA), corresponding to higher denudation rates.*

Low denudation rates measured at the outlet of the Rio Grande catchment are interpreted to be a result of a biased cosmogenic nuclide mixing that is dominated by headwater signals from the Eastern Cordillera and the Interandean zone and limited catchment sediment connectivity in the lower river reaches.

[1] F. Kober et al., Tectonophysics 657 (2015) 230

[1] *Geology, ETH Zurich*
[2] *Institute of Earth and Environmental Science, University of Potsdam, Germany*
[3] *Instituto de Hidraulica e Hidrologia, Universidad Mayor de San Andrés, Bolivia*

UPLIFT RATE DISTRIBUTION IN THE ANDES (~32° S)

Surface exposure dating with ^{10}Be

C. Terrizzano[123], E. García Morabito[123], R. Zech[123], Y. Yamin[4], N. Haghipour[3], L.Wüthrich[123], M. Christl

Understanding the deformation associated with active thrust wedges is essential to evaluate seismic hazard. In the central Andean backarc, for instance, controversy exists over how deformation is distributed in a millennial scale. To address this issue, we combined a structural and geomorphological approach with surface exposure dating (^{10}Be) of alluvial fans at ~32° S located both in the active thrust front and in the Andean interior largely considered to be a stable region (Fig. 1).

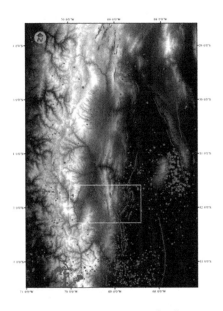

Fig. 1: *The study zone (yellow rectangle) in the Central Andes. Red lines: faults with Quaternary activity; Blue and purple points: <33 km earthquakes, M>5 and M<5 respectively [1].*

22 surface samples (boulders or pebbles or sand) and six depth profiles on sand were sampled in these two localities. Under the assumption of negligible erosion, ages of 100-130 ka were obtained for the oldest terrace (T1), 40-95 ka from the intermediate (T2) and ~20 ka from the youngest (T3) at the thrust front. In the Andean interior (Fig. 2) T1´ yields ages of 117-146 ka, T2´ is ~70 old and T3´ has an age of ~20 ka. Vertical slip rates of 0.3-0.4 mm/yr and of 0.6-1.2 mm/yr derived from the combination of fault offsets and ^{10}Be ages in the thrust front and in the Andean interior, respectively.

Fig. 2: *Alluvial terraces in the Andean interior.*

We argue that the deformation rates in the Andean interior are comparable with those along the Andean thrust front. This particular behaviour is most likely related to the reactivation of Paleozoic and Triassic structural heterogeneities and should be taken into account in terms of the direct impact in the welfare and safety of this populated area of Argentina.

[1]   U.S. Geological Survey earthquake data base

---

[1] *Geography, University of Bern*
[2] *Oeschger Centre, University of Bern*
[3] *Geology, ETH Zurich*
[4] *Segemar, Argentina*

CLIMATE CONTROL ON ALPINE HILLSLOPE EROSION

Using ^{10}Be in lake cores to reconstruct Holocene denudation rates

R. Grischott[1], F. Kober[2], M. Lupker[1], J.M. Reitner[3], R. Drescher-Schneider[4], I. Hajdas, M. Christl, S. Willett[1]

Climate is a major driver of landscape denudation but disentangling its effects from that of tectonics or anthropogenic remains challenging. We reconstruct paleo-denudation rates over Holocene timescales to isolate and test the climatic forcing on denudation in an Alpine catchment [1]. To overcome potential biases associated with sediment accumulation rates as proxies for basin wide denudation fluxes, we measured cosmogenic ^{10}Be on two sediment cores [2] from Lake Stappitz (Austrian Alps, Fig. 1) to derive a 10-kyr long paleo-denudation record of the upstream Seebach alpine valley. This record was combined with a two-year time series of denudation rates in the active stream Seebach.

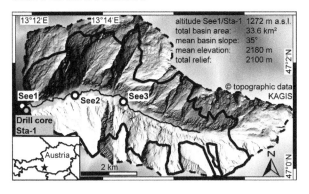

Fig. 1: *Outline of the study area (dashed) with the extent of the connected catchment (bold).*

While the record suggests a significant mixing with low-dosed glacial sediments from 10-8 kyr BP, there is a change in apparent denudation rate by a factor 2 during the period between 8 kyr BP and present that is attributed to the hillslope response to climate forcing (Fig. 2). Low hillslope erosion rates of ca. 0.4 mm/yr found during 5-8 kyr BP correlate with a stable climate, infrequent flood events and higher temperatures that favoured the widespread growth of stabilizing soils and vegetation. High hillslope erosion rates of ca. 0.8 mm/yr for the

last ~4 kyr correlate with an oscillating, cooler climate where frequent flood events increase denudation on the less protected hillslopes.

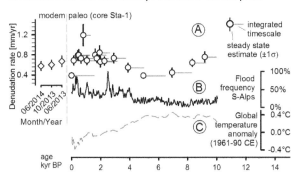

Fig. 2: *Holocene denudation rates from the core and the active stream correlate with flood frequency [3] and global temperature [4].*

Overall our results suggest a tight coupling of climate and hillslope erosion in Alpine landscapes. The results highlight the impact of transient climatic fluctuation on geomorphic processes and have important implications for the widespread use of cosmogenic nuclides as denudation proxy.

[1] R. Grischott et al., Geology (submitted)
[2] A. Fritz and F. Ucik, Naturwiss. Verein für Kärnten: Klagenfurt (2001)
[3] S.B. Wirth et al., Quat. Sci. Rev. 80 (2013) 112
[4] S.A. Marcott et al., Science 339 (2013) 1198

[1] Geological Institute, ETH Zurich
[2] NAGRA, Wettingen
[3] Geological Survey of Austria, Vienna
[4] Plant Sciences, K.-F.-University, Graz, Austria

FROST CRACKING AS DRIVER OF HOLOCENE EROSION

High resolution denudation rates from sediment cores of alluvial fans

R. Grischott[1], F. Kober[2], M. Lupker[1], K. Hippe, S. Ivy-Ochs, I. Hajdas, B. Salcher[3], M. Christl

[10]Be catchment-wide denudation rates (CWDR) that represent hillslope erosion have been widely used in landscape evolution studies. However, little is known about the variability of CWDRs and the potential impact of climatic fluctuations on Alpine denudation. Here, we present a 6 kyr long record of [10]Be paleo-CWDRs retrieved from sediment cores and a three-year time series from the active stream in the Fedoz Valley (Eastern Switzerland).

Paleo-CWDRs decrease with time from 1.2 mm/yr at 6 kyr BP to 0.6 mm/yr at present [1]. The data correlate with a record of global temperature variation [3] but not with flood frequency [2] or glacier fluctuations [4] likely due to a missing sensitivity for perturbations on sediment transport and mixing.

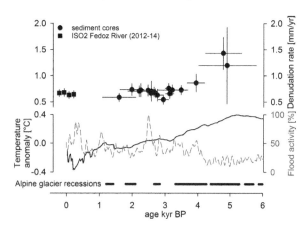

Fig. 1: *Paleo-CWDRs from the core and active stream plotted with flood frequency [2], global temperature [3] and glacier fluctuations [4].*

Results can be further correlated with a climatic model [5] to analyse the variations of the frost-cracking intensity and thus the potential variability of sediment production. Frost-cracking might explain the higher denudation rate observed in the Middle Holocene as the 0°C isotherm was significantly elevated and a larger

part of the catchment was affected by erosion through frost/thaw cyclicity.

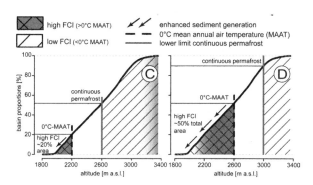

Fig. 2: *The two temporal endmember models of frost cracking for the Middle Holocene and today.*

Our data confirm previous findings that suggest a strong dependency of erosion on altitude-related geomorphic processes. They further highlight the limitations for global predictions of climate-induced denudation rate changes given by the complex responses of the landscape.

[1] R. Grischott et al., Earth Surf. Process. Landf. (submitted)

[2] S.B. Wirth et al., Quat. Sci. Rev. 80 (2013) 112

[3] S.A. Marcott et al., Science 339 (2013) 1198

[4] U.E. Joerin et al., Holocene 16 (2006) 697

[5] T.C. Hales and J.J. Roering, Geology 33 (2005) 701

[1] Geological Institute, ETH Zurich
[2] NAGRA, Wettingen
[3] Geological Survey of Austria, Vienna

THE IMPACT OF TYPHOON MORAKOT ON EROSION RATE

Temporal changes in erosion rate in SOUTHERN TAIWAN

C.Y. Chen[1], S.D. Willett[1], A.J. West[2], S. Dadson[3], M. Christl

Super-typhoon Morakot impacted Taiwan in 2009, bring a large amount of rainfall and triggering several thousand landslides in southern Taiwan. We obtained two sample sets, one in 2006, one in 2012/2014, bracketing this event, to measure its impact on ^{10}Be concentrations and the inferred erosion rate.

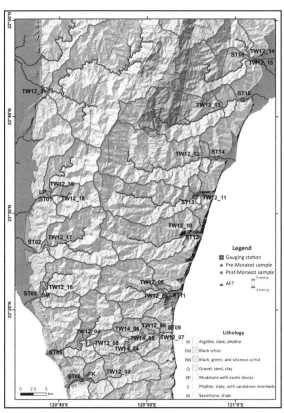

Fig. 1 *Geological map and sample distribution. Blue and red dots are pre-Morakot and post-Morakot samples, respectively. AFT-derived erosion rates are calculated from data published in [1] (using method of [2]).*

The spatial pattern of basin-wide erosion rate correlates with topography, the average basin steepness, and the AFT-derived erosion rate, all showing a gradual increase to the north. Besides, post-Morakot samples yield higher inferred erosion rates, suggesting dilution of the

sediment by quartz with low concentrations of ^{10}Be from the deep-seated landslides triggered by Morakot.

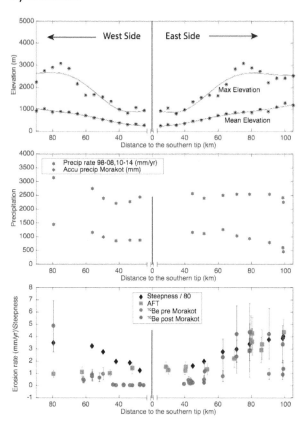

Fig. 2 *Comparison of topography with precipitation, average basin steepness and erosion rates.*

[1] Fuller et al., Tectonophysics 425 (2006) 1
[2] Willett and Brandon, G^3 14 (2013) 209

[1] *Geological Institute, ETH-Zurich, Switzerland*
[2] *Department of Earth Sciences, USC, USA*
[3] *School of Geography and the Environment, University of Oxford, UK*

ANTHROPOGENIC RADIONUCLIDES

The ^{236}U input function for the NE-Atlantic

TRANSARC II: Expedition to the Arctic Ocean

The geotraces cruise GA01

First ^{236}U data from the equatorial Pacific Ocean

^{129}I off the coast of Japan

Artificial radionuclides off the coast of Japan

Determination of Pu and U in Japanese samples

Radionuclides in drinking water reservoirs

THE ^{236}U INPUT FUNCTION FOR THE NE-ATLANTIC

A first step towards ^{129}I/^{236}U and ^{236}U/^{238}U based tracer ages

M. Christl, N. Casacuberta, C. Vockenhuber, C. Elsässer[1], P. Bailly du Bois[2], J. Herrmann[3], H.A. Synal

Anthropogenic ^{236}U and, to a minor extend, also ^{129}I have been introduced into the world oceans as a result of atmospheric nuclear bomb tests. Additionally, both radionuclides are discharged into the Northeast Atlantic Ocean by the two nuclear reprocessing facilities Sellafield (SF) and La Hague (LH, Figure 1).

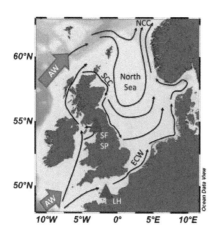

Fig. 1: *Map of the North Sea region showing the major currents and water masses (arrows) together with the location of the nuclear facilities (triangles).*

In the North Sea region Atlantic Waters (AW) carrying the bomb fallout signature of ^{129}I and ^{236}U are mixing with English Channel Waters (ECW) and with waters of the Scottish Coastal Current (SCC) that additionally carry the radionuclide signatures of LH and SF (and potentially also of the nuclear fuel producing facility Springfields, SP).

In our recent study [1] a first reconstruction of the ^{236}U input function for the Northeast Atlantic Ocean has been constructed and, in combination with ^{129}I, the input functions of ^{129}I/^{236}U and ^{236}U/^{238}U are presented (Figure 2). Our results show that, since about 1990, the ^{129}I/^{236}U input function steadily rises. This implies that the ^{129}I/^{236}U ratio can be used over the past about 25 yr to estimate tracer ages or

transit times of AW that have passed by the North Sea region before entering the Arctic Ocean via the Norwegian Coastal Current (NCC, Figure 1).

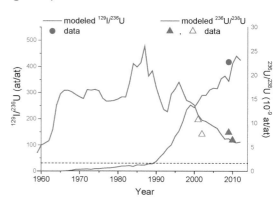

Fig. 2: *Reconstruction of the ^{129}I/^{236}U and ^{236}U/^{238}U (both at/at) input functions for the North Sea region. Data points represent measured values (median data) for the respective year.*

A first comparison with ^{129}I and ^{236}U data from the Arctic Ocean (sampled in 2011/12) shows that the ^{129}I/^{236}U and ^{236}U/^{238}U based tracer age concept works well if dilution effects (i.e the ongoing mixing with ^{129}I and ^{236}U from global fallout) are taken into account. Our dilution corrected tracer ages correspond well with independently derived values for the Arctic Ocean. This shows that ^{129}I/^{236}U can be used as a valuable new tool in tracer oceanography not only for the determination of transit times in the Arctic Ocean.

[1]   M. Christl et al., JGR Oc. 120 (2015) 7282

---
[1] *Environmental Physics, Univ. Heidelberg, Germany*
[2] *IRSN/PRP, Cherbourg-Octeville, France*
[3] *BSH, Hamburg, Germany*

TRANSARC II: EXPEDITION TO THE ARCTIC OCEAN

Sampling for artificial radionuclides on board Polarstern

N. Casacuberta, M. Christl, C. Vockenhuber, T. Kenna[1], M. R. van-der-Loeff[2]

The TransArc II expedition consisted of 55 scientists and 43 crew members sailing for two months (17th August - 15th October) to the Arctic Ocean on the German research ice breaker Polarstern. The aim was to track the climate change, which is particularly prominent in the Arctic Ocean e.g. with its drastic sea ice reduction. The cruise was part of the International Arctic GEOTRACES field program for 2015, together with other two expeditions from Canada (on board CCGS Amundsen) and one from USA (on board USCG Healy). At two crossover stations samples were taken by all counterparts to ensure quality and accuracy of the Arctic-wide observations.

Fig. 1: *R/V Polarstern during the TransArc II.*

Many parameters were sampled in 33 stations during the expedition, with the aim to determine distributions of selected trace elements and isotopes (TEIs), including their concentration, chemical speciation and physical form, and to evaluate the sources, sinks and internal cycling of these species.

Other than TEIs, transient tracers such as CFCs, ^{3}He, SF_6 and artificial radionuclides can provide information on circulation pathways and time scales, mixing, time of isolation of the surface water beneath the ice cover and inflow of dense shelf waters to the deep basins [1]. Our role on the TransArc II expedition was sampling for

artificial radionuclides: ^{129}I, ^{236}U and Pu-isotopes. More than 150 samples were collected for each radionuclide along the transect covering the Barents Sea, the Eurasian Basin and crossing the North Pole to the Makarov Basin.

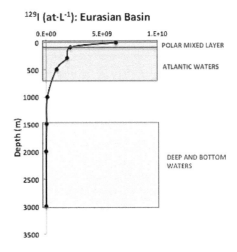

Fig. 2: *Profile of ^{129}I in the Eurasian Basin.*

Preliminary results of ^{129}I concentrations (Fig. 1) in the Eurasian Basin show the inflow of the Atlantic waters to the Arctic Ocean in the upper 1000 m layer. These waters carry the signal of ^{129}I from the two reprocessing plants of Sellafield and La Hague. Deep waters are more isolated and thus older, resulting in much lower concentrations of ^{129}I. A quasi-synoptic full transect of ^{129}I in the Arctic Ocean will be produced after gathering the results of the four Arctic expeditions.

[1] H. D. Livingston et al, Health Phys. 82 (2001) 656.

[1] *Lamont-Doherty Earth Observatory, USA.*
[2] *Alfred Wegener Institute, Germany.*

THE GEOTRACES CRUISE GA01

New section for ^{129}I and ^{236}U in the North Atlantic Ocean

M. Castrillejo[1], P. Masqué[1], Jordi Garcia-Orellana[1], N. Casacuberta, M. Chirstl, C. Vockenhuber

The North Atlantic Ocean is crucial for the Earth's climate as it represents the major overturning area of the global thermohaline circulation, the so-called Atlantic Meridional Overturning Circulation (AMOC). The net result of the AMOC is the formation of dense water of Atlantic and Arctic origin, which then transport heat, momentum and geochemical elements from surface to abyssal depths. Some of these elements (e.g. carbon) are directly involved in the cycling of atmospheric CO_2 or in the primary production (e.g. Fe, Zn, Co), while others (e.g., artificial radionuclides) can be used as tracers of circulation and particle scavenging.

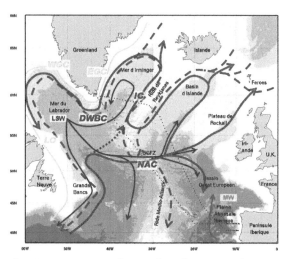

Fig. 1: *Mean surface distribution of major currents and water masses in the North Atlantic.*

The Geovide cruise (Geotraces GA01 section, Figure 2) aimed to better know and quantify the MOC and the carbon cycle, using new key tracers. ^{236}U has recently emerged as a potential new tracer of ocean circulation (e.g. [1]) and its combined use in a dual tracer approach (^{129}I/^{236}U and ^{236}U/^{238}U) provides means of calculating tracer ages and ventilation rates in the North Atlantic [2]. Geovide benefits from existing ^{236}U data for the Western North Atlantic

Ocean [3] and a strong physical oceanographic background acquired through the OVIDE project (IFREMER, 2002-2012).

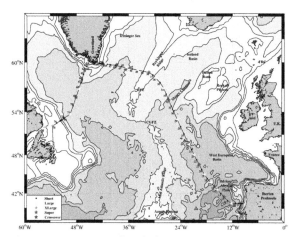

Fig. 2: *Cruise track of the Geovide cruise. Red dots indicate sampled locations.*

Over 150 seawater samples for ^{236}U and ^{129}I were collected on board *R/V 'Pourquoi Pas?'* between Lisbon, Portugal, and St. John's, Canada, in May-June 2014 (Figure 2). 14 vertical profiles and the surface ocean were sampled obtaining a full-depth high resolution along the section. After radiochemical purification, ^{236}U and ^{129}I will be measured using the compact Tandy AMS system at ETH-Zürich. The new dual tracer approach shall be tested with the new data within the AMOC region.

[1] M. Christl et al., Geochim. Et Cosmochim. Acta 77 (2012) 98.

[2] M. Christl et al., J. Geophys. Res. Oc. 120 (2015) 7282.

[3] N. Casacuberta et al., Geochim. Et Cosmochim. Acta 133 (2014) 34.

[1] *Env. Sci. and Phys., Aut. Univ. of Barcelona, Spain*

FIRST ^{236}U DATA FROM THE EQUATORIAL PACIFIC OCEAN

Measurements from the GEOTRACES Equatorial Pacific Zonal Transect

E. Chamizo[1], M. Villa[2], , M. López-Lora[1], N. Casacuberta, M. Christl, T. Kenna[3]

The U.S. GEOTRACES East Pacific Zonal Transect (EPZT) was sampled in October to December 2013. The transect from Peru to Tahiti included the Peru Margin, upwelling and oxygen minimum zone (OMZ), St. 11 (12°S,94°W), and the large hydrothermal plume (HP), St. 18 (15°S,113°W), originating from the southern East Pacific Rise (Fig. 1).

In total 125 4L samples were collected, acidified and spiked with ^{233}U. Actinides were pre-concentrated at Lamont-Doherty Earth Observatory (LDEO) and ^{236}U radiochemical separation was performed at CNA. Expected ^{236}U/^{238}U atom ratios in surface samples (above 400 m) were at the order of 10^{-10}. These samples were measured on the 1 MV CNA AMS system, which provides a ^{236}U/^{238}U background ratio of 7×10^{-11}. Deeper samples with estimated ^{236}U/^{238}U ratios below 10^{-10} were measured at the ETH Tandy facility, that provides a background level below 10^{-13}.

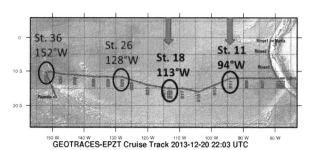

Fig. 1: *GEOTRACES East Pacific Zonal Transect*

The results (Fig. 2) show that: a) Most of the samples below 600 m show very low ^{236}U/^{238}U ratios at the 10^{-12} level, which is close to the estimated lithogenic/natural background. These ratios are the lowest values measured so far at the ETH facility. b) Samples above 600 m show ratios around 10^{-10}, similar to values found in the shallow Equatorial North Atlantic (EqNA) [1] and clearly indicating the presence of anthropogenic ^{236}U. c) Similar ^{236}U/^{238}U ratios were found in the upper 1300 m in both, the OMZ and the HP. Below, increased ^{236}U levels are found towards the sediment interface at the HP station, and a significant variation of ^{236}U with depth is found in the OMZ profile. The analysis of ^{236}U at St. 26 and 36 (Fig. 1) will provide further information to explain the observed behavior.

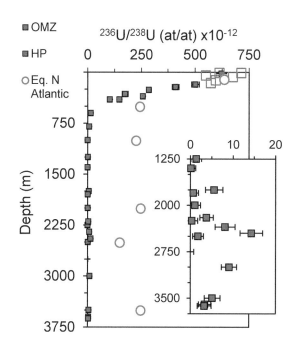

Fig. 2: ^{236}U/^{238}U in St. 11 (OMZ), St.18 (HP) and the EqNA (2.54°N 41.7°W). Hollow squares were samples measured at CNA

[1]   N. Casacuberta et al., Geochim. Cosmoschim. Acta 133 (2014) 34.

---
[1] *Centro Nacional de Aceleradores (CNA), Sevilla University, Spain.*
[2] *Applied Physics Dep. Sevilla University, Spain.*
[3] *Lamont-Doherty Earth Observatory, University of Columbia, USA.*

^{129}I OFF THE COAST OF JAPAN

Continuous releases of ^{129}I from Fukushima in 2013 and 2014

Y.S. Lau[1], N. Casacuberta, C. Vockenhuber, M. Christl, J. Pates[1]

The nuclear Fukushima Daiichi Nuclear Power Plant (1FNPP) accident that occurred in March 2011, resulted in a significant release of radionuclides to the environment. The broad suite of radionuclides emitted from the FNPP was released via atmospheric plumes and direct discharge into the nearby ocean. The releases also included the iodine radioisotopes ^{131}I and ^{129}I. However, due to the short half-life of ^{131}I ($T_{1/2}$ = 8 days), only ^{129}I ($T_{1/2}$ = 15.7 Myr) can be measured today either to retrospectively infer ^{131}I, or as a oceanographic tracer in the Pacific Ocean.

Sources of ^{129}I to the marine environment are: 1) atmospheric fallout of nuclear bomb testing, 2) discharge from nuclear fuel reprocessing plant and 3) nuclear accidents. In the case of 1FNPP accident, ^{129}I can also serve as an indicator of polluted water discharge. The aim of this study was to determine the concentrations of ^{129}I at the east coast of Japan 2-3 years after the accident, to test for potential releases of this isotope. For this purpose, three sets of iodine samples (n_{total} = 56) were collected in 2013 and 2014 during three different expeditions (Sept 2013, May 2014 and Oct 2014). The radiochemical separation of ^{129}I was carried out according to [1], sample analysis was performed by AMS using the compact ETH TANDY system.

Results show high concentration of ^{129}I (778±13 x10^7 at·kg^{-1}) at the closest station to 1FNPP (Figure 1). Compared with the pre-Fukushima levels of ^{129}I in this area (about 1x10^7 at·kg^{-1}[2]), the above value provides strong evidence for a discharge of contaminated water from 1FNPP in September 2013. Similar results were obtained for samples in 2014, with concentrations of ^{129}I up to (21.0±0.4) x10^7 at·kg^{-1} at the same station as in 2013.

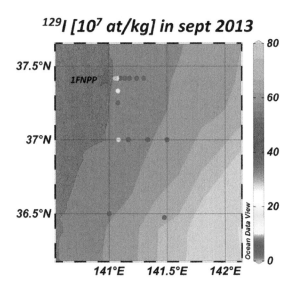

Fig. 1: *Concentration of ^{129}I in surface water samples around 1FNPP in September 2013.*

It has been stated in many studies (e.g. [3]) that the distribution of 1FNPP-derived radionuclides is strongly affected by the Oyashio and Kuroshio Currents. The spatial distribution of our results also reflects the transport of ^{129}I from North to South driven by the Oyashio current. However, even if the 1FNPP-derived ^{129}I is clearly detectable on a local scale, the signal is rapidly diluted while spreading into the Pacific Ocean. Thus, the clear impact of ^{129}I that is found around the coast off Fukushima is negligible on a larger, ocean basin wide scale.

[1]   R. Michel et al., Sci. Tot. Env. (2012) 419
[2]   T. Suzuki et al., Biogeosciences. (2013) 10
[3]   K. Buesseler et al., Envir. Sci. Tech. (2011) 45

[1] University of Lancaster, UK.

ARTIFICIAL RADIONUCLIDES OFF THE COAST OF JAPAN

Tracking radioactive releases 2-3 years after the Fukushima accident

N. Casacuberta, M. Christl, C. Vockenhuber, Y.S. Lau[1], P. Masqué[2], K.O. Buesseler[3]

Four different cruises took place off the coast of Japan during 2013, 2014 and recently one in October 2015, with the aim to understand the sources, fate, transport and associated impact of radionuclides from Fukushima. These cruises took place within the remit of the EU FRAME project.

All four cruises collected surface and shallow depth profile samples from the same stations in order to have a time evolution within the same domain (Fig. 1). Samples for ^{137}Cs, ^{134}Cs and ^{90}Sr analysis were processed and measured at WHOI [1]. Long-lived radionuclides (^{129}I, ^{236}U and Pu-isotopes) were analyzed at ETH Zürich.

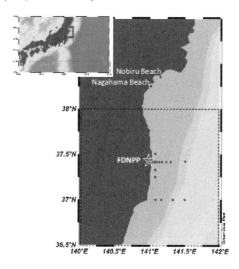

Fig. 1: *Map of the stations sampled during the cruises in 2013 and 2014. The black dashed square indicates the domain in Figure 2.*

The presence of relatively high concentrations of ^{129}I (up to 800 x10^7 at·kg^{-1}) in the waters close to the Fukushima Dai-ichi Nuclear Power Plant (FDNPP) indicates releases of this radionuclide in the years after the accident [2].

Concentrations of ^{236}U, sampled in 2013 and 2014 (Fig. 2), do not show a significant increase from values expected from global fallout.

Indeed, concentrations are similar to the ones found in the western North Atlantic Ocean at similar latitude [3].

^{240}Pu/^{239}Pu atomic ratios range from 0.21±0.01 to 0.29±0.03, with all ratios falling within the global/regional fallout range for the Pacific Ocean (Fig. 2). Yet the highest ^{240}Pu/^{239}Pu atomic ratio that also corresponds to the highest ^{240}Pu and ^{239}Pu concentrations was measured close to the FDNPP. This might indicate a potential release of Pu-isotopes to the coast off Japan.

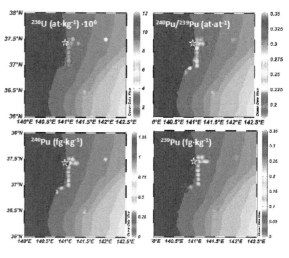

Fig. 2: *^{236}U concentrations, ^{240}Pu/^{239}Pu atomic ratios and ^{240}Pu and ^{239}Pu concentrations from samples taken in October 2014.*

[1] M. Castrillejo et al., Environ. Sci. Technol. (2015) in press

[2] Y.S. Lau et al., Annual Report LIP (2015)

[3] N. Casacuberta et al., Geochim. Cosmochim. Acta 133 (2014) 34

[1] *University of Lancaster, UK.*
[2] *The University of Western Australia, Australia.*
[3] *Woods Hole Oceanographic Institution, USA.*

DETERMINATION OF PU AND U IN JAPANESE SAMPLES

Actinide investigations in environmental samples from Fukushima

S. Schneider[1], M. Christl, K. Shozugawa[2], G. Steinhauser[1], C. Walther[1]

During the accident of the power plant Fukushima Daiichi only a small amount of actinides were released into the environment. Previous investigations showed a strong localization of plutonium [1]. Because of the high radiological risk if incorporated, further investigations were necessary.

In June 2013 vegetation, litter and soil drill core samples were taken at different sites in the vicinity of the damaged power plant. The sampling sites are shown in Figure 1. The core samples were taken in a depth up to 12 cm, each core was split into six equal segments. After a chemical treatment, all samples were investigated using accelerator mass spectrometry to determine the isotopic ratios Pu^{240}/Pu^{239} and U^{236}/U^{238}. These ratios are used as an indicator of the origin of the current nuclides, so it is possible to distinguish between global fallout and material of the reactor.

Fig. 1: *Sampling sites in Fukushima*

In total 64 samples were measured and most samples show no significant deviation from values of global fallout of Pu and the natural distribution of U. In Table 1 the values which differ most strongly are shown for both elements.

Sample	$^{240}Pu/^{239}Pu$	$^{236}U/^{238}U$
Lit-C	0.29 ± 0.04	---
Lit4 – D	0.29 ± 0.02	---
Veg – F	0.32 ± 0.05	---
F1-31 3'' – H	0.48 ± 0.15	---
F1-06 3'' – B	---	$1.37 * 10^{-7}$

Tab. 1: *Results of the most relevant samples*

Only 4 samples show ratios indicating a contamination with plutonium from the Fukushima reactors. In most cases the plutonium is located in the upper layers like in vegetation or litter. Only in sample F1-31 3''– H the ratio is very high in the third layer of the soil core, which is absolutely unexpected and needs further investigations.

In most samples ^{236}U is present in a natural distribution. Only in one sample (F1-06 3'' – B) the ratio is higher, which theoretically indicates an anthropogenic influence. However, such ratios were also measured in Japan before the accident happened [2].

For further investigations new samples were taken at nearly the same places in May 2015. They are processed and will be measured by AMS for their Pu and U ratios in the near future

[1] S. Schneider et al., Sci. Rep. 3 (2013) 2988
[2] A. Sakaguchi et al., Sci. Total Environ. 407 (2009) 4238

[1] Radioecology and Radiation Protection, University of Hannover, Germany
[2] Graduate School of Arts and Sciences, The University of Tokyo, Japan

RADIONUCLIDES IN DRINKING WATER RESERVOIRS

Sensitivity of reservoirs to input of man-made radionuclides

B. Riebe[1], S. Bister[1], A.A.A. Osman[1], A. Daraoui[1], C. Walther[1], C. Vockenhuber, H-A. Synal

Man-made radionuclides from nuclear facilities are deposited by precipitation, transported through the soil by percolating water, and are possibly carried to the groundwater. In a current project we are aiming to assess the sensitivity of an unconfined aquifer in Northern Germany (Fuhrberger Feld), which is used as a drinking water reservoir, with regard to introduction and accumulation of radionuclides, e.g. ^{129}I. Water samples are drawn from different depth (4 m, 14 m) by multilevel wells (Fig. 1) aligned in the direction of groundwater flow. A detailed description of the sample preparation is given elsewhere [1]. For the analysis of ^{129}I accelerator mass spectrometry (AMS) is used, ^{127}I concentrations are determined by inductively coupled plasma mass spectrometry (ICP-MS).

Fig. 1: *Drawing groundwater samples from a multilevel well.*

First results reveal that ^{129}I concentrations as well as ^{129}I/^{127}I ratios in the water samples from Fuhrberger Feld are about one order of magnitude higher than those from water samples from other aquifers from Lower Saxony (Fig 2.). For Fuhrberger Feld, concentrations of ^{129}I in the ground water samples varied between 2.1×10^{-14} and 8.0×10^{-14} g kg^{-1}, and therefore lie within the range of values determined for surface water samples from Lower Saxony. In comparison, ^{129}I concentrations measured for samples from confined aquifers from the same region were as low as 1.3×10^{-15} to 1.1×10^{-14}.

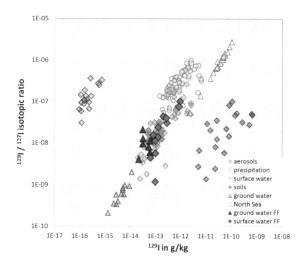

Fig. 2: : ^{129}I/^{127}I isotopic ratios versus ^{129}I concentration for groundwater samples in comparison to other environmental compartments in Lower Saxony, and North Sea water [1].

The same is true for the ^{129}I/^{127}I ratio. Values of groundwater samples from Fuhrberger Feld varying between 4.1×10^{-9} and 1.5×10^{-8} were in the range of surface water samples from the same sampling area (1.2×10^{-9} to 2.0×10^{-7}), whereas measurements of samples from confined aquifers resulted in values, which were one order of magnitude lower. This indicates that ^{129}I from the atmosphere has already reached the Fuhrberger Feld aquifer.

[1]   R. Michel et al., Sci. Tot. Environ. 419 (2012), 151

[1] *Institute for Radioecology and Radiation Protection, Leibniz University Hannover, Germany*

MATERIALS SCIENCES

The startup and first results of MeV SIMS at ETHZ

Novel approach for time of flight MeV SIMS

Stoichiometry studies in pulsed laser deposition

Hydrogen intake during atomic layer deposition

Monolayer detection of bipyridine

Composition uniformity of ablation targets

Production of a thin V-48 positron source

THE STARTUP AND FIRST RESULTS OF MEV SIMS AT ETHZ

Progress of the CHIMP setup

M. Schulte-Borchers, M. Döbeli, A.M. Müller, M. George, H.-A. Synal

The new **C**apillary **H**eavy **I**on **M**eV-SIMS **P**robe (**CHIMP**) set-up was completed early this year with attachment of a new gas detector in transmission geometry. So far it features a channeltron secondary electron detector and a micro channel plate detector for secondary ion detection at the end of the almost 0.5 m long time of flight (ToF) mass spectrometer. The primary beam is currently collimated to 1 mm diameter using an aperture. In the future this aperture will be replaced by a glass capillary for micrometer diameter beams to enable imaging of sample surfaces. Positioners for sample scanning and adjustment of the capillary angle are already included and tested in the chamber.

Fig. 1: Inside of the new CHIMP chamber. The sample is in the center (1), electron detector (2) and ToF extraction (3) are in front. Primary beam enters from the top (4) and the gas detector is at the bottom (5). To the side, entry lock and sample magazine (6) are indicated.

Recently, several system parameters have been evaluated and optimized. A mechanical adjustment of the sample holder to decrease the distance from sample to ion extraction nozzle proved necessary for better extraction.

The setup is designed for multiple measurement modes of the ion flight time: Either pulsing of the primary beam or secondary electrons from the sample can be used to start the flight time measurement. Additionally, transmitted ions can be detected with the new gas detector for thin samples. These modes have been compared to achieve optimal parameters. Best results in terms of mass resolution were achieved with the electron detector (see Fig. 2), which is now performing with good efficiency and time resolution [1].

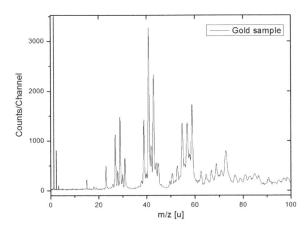

Fig. 2: Positive secondary ion mass spectrum of a solid gold sample measured with 28 MeV Au^{6+} primary ions. Several contamination peaks can be identified on the uncleaned surface.

In this case, a time resolution under 10 ns (FWHM) and extraction efficiencies of several percent were measured. As a next step the imaging properties of the set-up will be investigated.

[1]   M. Schulte-Borchers, LIP Annual Report (2015) 89

NOVEL APPROACH FOR TIME OF FLIGHT MEV SIMS

Detecting primary ion hits by secondary electrons

M. Schulte-Borchers, A.M. Müller, M. George, M. Döbeli, H.-A. Synal

Typical MeV SIMS setups with Time-of-Flight (ToF) mass spectrometers use a pulsed or chopped primary ion beam. This provides a well-defined ToF start signal from the controls of the pulsing. Time resolution is then determined mainly by the length of the primary ion pulse, which is in the order of tens of nanoseconds [1] if state-of-the art fast high voltage switches are used.

The main disadvantage of such a system is the enormous decrease of duty cycle, as most of the time the beam is not on the sample. For the capillary based MeV SIMS imaging setup at LIP a high duty cycle measurement mode is needed. Capillary collimation reduces the initial beam current by a factor of about one million; an additional reduction by several orders of magnitude due to pulsing would prohibit measurements with rare beam particles such as high energy cluster ions.

It was therefore decided to use secondary electrons generated upon each primary ion impact on the sample as a fast and precise ToF start signal (Fig. 1).

By carefully adjusting the positive and negative voltages for extraction of electrons and positive ions to either side of the sample it is possible to obtain simultaneous signals from both detection systems. Fig. 2 shows a comparison of the positive ion mass spectrum obtained from a teflon sample with the existing AMS beam pulsing system and with the novel secondary electron start signal. The result is a major improvement in resolution. While the smallest observed ToF peak width with beam pulsing is about 160 ns it is less than 10 ns (FWHM) with secondary electrons. The duty cycle was increased by approximately 4 orders of magnitude.

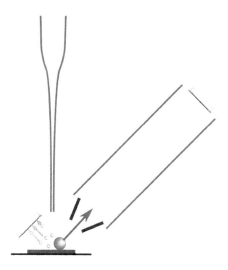

Fig. 1: *Sketch of the sample zone in the MeV SIMS setup. The primary beam is collimated with a capillary onto the sample. Secondary electrons and ions are extracted to the according detection systems.*

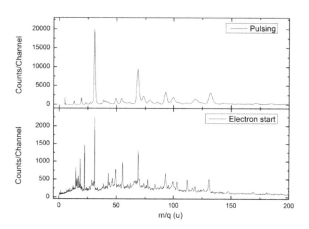

Fig. 2: *Positive secondary ion mass spectra of PTFE taken with beam pulsing and electron start signal.*

[1]   T. Tadic et al., Nucl. Instrum. Meth. B 332 (2014) 234

STOICHIOMETRY STUDIES IN PULSED LASER DEPOSITION

Angular dependence of thin copper-gold film composition

Stela Canulescu [1], Jørgen Schou[1], M. Döbeli

Thin films of Cu-Au alloys were deposited in vacuum and background gas by Pulsed Laser Deposition (PLD) to investigate the composition of films deposited at different angles around the expanding plume. The target containing light and heavy atoms at a comparable ratio (50 at% Cu, 50 at% Au) was irradiated with a Nd:YAG laser at a wavelength of 355 nm. The ablated material was collected on Si substrates placed over a wide range of angles (Fig. 1).

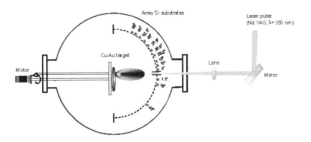

Fig. 1: *Setup used for deposition of Cu-Au metallic films by PLD. The laser beam is incident onto the target with Si substrates of 5x5 mm^2 size placed over a full hemispherical array.*

The composition of the Au-Cu films was analyzed by 2 MeV He RBS. At relatively large laser fluence (5 J/cm^2) the ratio of Cu to Au atoms deviates significantly from the target stoichiometry, with large depletion of the lighter component (Cu) at angles close to normal incidence (Fig. 2). This may indicate that the light species (Cu) is scattered off the plume, leading to a significant excess of Au in the deposited films.

Films produced by PLD contain large size droplets which arise from rapid melting of the subsurface layer under high laser power irradiation. RBS with 5 MeV He was therefore used to investigate the composition of films including droplets (also included in Fig. 2). A comparison between composition of films deposited in vacuum and Xe background gas is shown in Fig. 3.

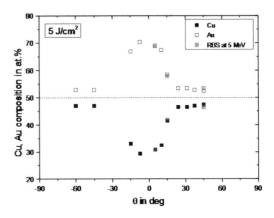

Fig. 2: *RBS analysis of Au-Cu films deposited in vacuum at different angles. The dashed line is the estimated ablation target composition.*

It is evident that presence of Xe enhances scattering of Cu atoms in the plume, leading to a much broader distribution of Cu and Au atoms with respect to vacuum.

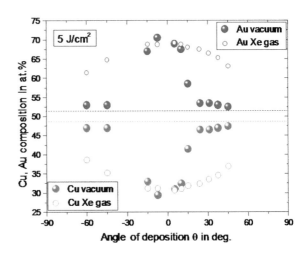

Fig. 3: *RBS analysis of Au-Cu films deposited in vacuum (full spheres) and 0.1 mbar Xe gas (empty circles) at different angles.*

---

[1] *Photonics Engineering, Technical University of Denmark, Denmark*

HYDROGEN INTAKE DURING ATOMIC LAYER DEPOSITION
HE ERD STUDY WITH DEUTERIUM TRACER

I. Utke[1], C. Guerra-Nunez[1], L. Pillatsch[1], M. Döbeli

In Atomic Layer Deposition (ALD) a thin film is grown by sequential exposure of a surface to typically two precursor gases (Fig. 1). Especially in low temperature processes the surface reaction can be incomplete and unwanted constituents of the precursors can be built into the film. In this project the incorporation of hydrogen into Al_2O_3, TiO_2 and ZnO via the precursors $Ti(iOPr)_4$ [1], $Al(CH_3)_3$, $Zn(Et)_2$ [2], and D_2O are investigated.

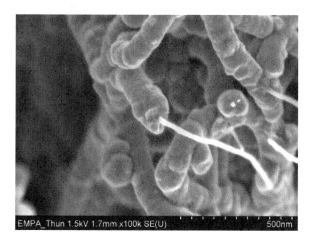

Fig. 1: *Uniform TiO₂ coating of a carbon nanotube array by ALD.*

In order to clarify the origin and the incorporation mechanism of the hydrogen a series of films have been grown at different deposition temperatures using deuterated water as oxidizing reactant. These samples have been characterized by 2 MeV [4]He RBS and 13 MeV [127]I Heavy Ion ERDA to determine the main oxide composition and by 2 MeV [4]He ERD with absorber foil to measure the quantitative hydrogen depth profile. For thin films H and D can be separated due to their large difference in kinematic factor and only a single ERD measurement is necessary. Fig. 2 shows an example for an aluminum oxide layer deposited at 40°C. The profiles show that the natural hydrogen mainly diffused in from the surface after the deposition while deuterium from the D_2O precursor was homogeneously built into the material during the growth process.

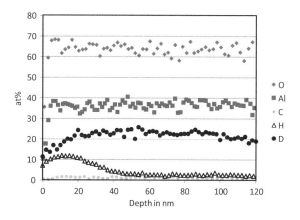

Fig. 2: *ERDA depth profile of an Al₂O₃ film deposited at 40° C. Atomic concentrations are normalized to Al + O = 100 %.*

The measurements reveal that for 40°C ALD of Al_2O_3, hydrogen is mainly built in from the oxidizer (about 20 at.%) and thus from unreacted OH groups while hydrogen incorporation through unreacted CH_x ligands of the metal precursors is at the 2 at.% level. These levels are high compared to TiO_2 and ZnO ALD and we presently investigate a correlation with the microstructure of the films employing XRD, XRR, SEM, and ToF-SIMS measurements.

[1]   C. Guerra et al., Nanoscale 7 (2015) 10622
[2]   R. Raghavan et al., Appl. Phys. Lett. 100 (2012) 191912

[1] *Laboratory for Mechanics of Materials and Nanostructures, EMPA Thun.*

MONOLAYER DETECTION OF BIPYRIDINE

Structure of a monomer monolayer at the air/water interface

W. Dai[1], B. Zelenay[1], T. Kepplinger[1], J. Sakamoto[1], A. Schütz[1], Z. Zheng[1], A. Borgschulte[2], L.-T. Lee[3], W. Abuillan[4], M. Tanaka[4,5], D. Schlüter[1], M. Döbeli[1]

Monolayer (ML) sheets with strong internal bonds in principle allow for rational structure design and placement of functional groups at predetermined sites. We have started to explore the accessibility of ML sheets including two-dimensional polymers at the air/water interface. A bipyridine based monomer (Fig. 1) was synthesized and investigated for possible metal complexation. A dilute chloroform solution of the monomer was spread at an air/water interface. The formation of a ML sheet was observed, was exposed to concentrated $Ni(ClO_4)_2$ metal salt solutions and horizontally transferred onto SiO_2 coated Si wafers.

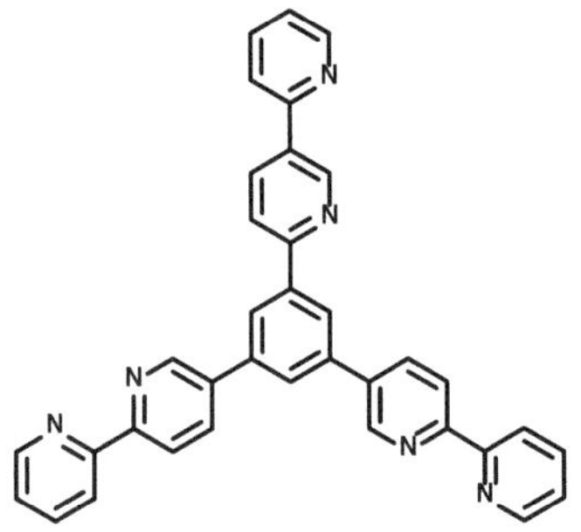

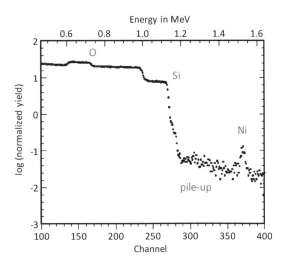

Fig. 2: *2 MeV ^{4}He RBS spectrum of the ML film. The Ni coverage is $(3.5 \pm 0.5)\cdot 10^{14}$ at/cm^2.*

Although the data analysis is still in progress, it can be concluded that Ni is part of all sheets and the Ni(II):monomer ratio is on the order of 1.

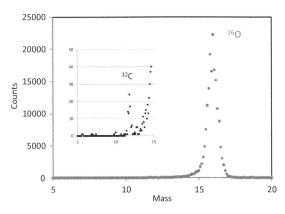

Fig. 1: *Structure of bipyridine-based monomer carrying three bipyridine units to serve as ligands for metal ion (e.g. Ni) complexation.*

The thickness of the Ni-exposed ML is 19 Å as determined by TM-AFM. We measured the Ni coverage by RBS (Fig. 2) and the C coverage by Heavy Ion ERDA (Fig. 3). For both techniques the detection of such small quantities is very challenging. Small count rates and long measurement times have been chosen to avoid background from pulse pile-up events. Together with results from XRR, NR, XPS and AFM thickness analysis, we will build a structure model for the Ni-exposed sheet.

Fig. 3: *ERDA mass spectrum of the ML film. The carbon coverage is $(6.6 \pm 0.8)\times 10^{14}$ at/cm^2.*

[1] *Department of Materials, ETH Zurich*
[2] *Empa Dübendorf*
[3] *Laboratoire Léon Brillouin, CEA-Saclay, France*
[4] *Physical Chemistry, Univ. of Heidelberg, Germany*
[5] *Cell Material Sciences, Univ. of Kyoto, Japan*

COMPOSITION UNIFORMITY OF ABLATION TARGETS

In-air micro-PIXE study of laser ablation targets used in PLD

A. Ojeda [1], T. Lippert [1,2], M. Schulte-Borchers, M. Döbeli

In Pulsed Laser Deposition (PLD) congruent transfer of the material composition from the ablation target to the deposited film is a crucial issue. In order to investigate this problem it is necessary to measure the exact composition of both, the deposited film and the ablation target. Even if the target is made of a well-defined material it is possible that the surface is altered during the ablation process and different material phases can be formed. It is therefore necessary to map the composition on the target surface on ablated and pristine locations. Since ablation targets are rather bulky and have complicated shapes remote handling in vacuum is problematic.

Fig. 1: *CaTiO₃ ablation target on the XY table of the in-air PIXE set-up. The beam collimating glass capillary enters from the left, the SDD X-ray detector is at the top left corner.*

On these grounds, 2 MeV proton PIXE measurements on $CaTiO_3$ and $La_{0.4}Ca_{0.6}MnO_3$ targets have been performed on the LIP in-air capillary microprobe [1]. A glass capillary with an outlet diameter of about 15 µm has been used having a wall thickness large enough to absorb the X-rays produced on the inner surface of the capillary. Therefore no extra shielding towards the SDD (Silicon Drift Diode) detector

was necessary (Fig. 1). Several line scans were carried out along the cylinder axis of the targets. Fig. 2 shows the ratio of the Ca and Ti K_α X-ray line intensities measured on the surface of the $CaTiO_3$ target along a straight line perpendicular to the well visible "zebra" pattern. This pattern is produced by the periodic linear and rotational movement of the target during laser ablation.

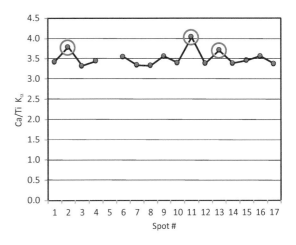

Fig. 2: *Ratio of Ca and Ti K_α X-ray intensities along a linear surface scan. Red circles mark positions of dark areas in the "zebra" pattern.*

The measurements reveal that the dark color change on the surface is probably due to a preferential loss of Ti in the strongly ablated areas.

The in-air capillary microprobe has proven to be a fast and simple tool to map elemental concentrations on objects that are difficult to manage in a vacuum chamber.

[1]   M.J. Simon et al., Nucl. Instr. and Meth. B273 (2012) 237

[1] *Materials Group, Paul Scherrer Institut*
[2] *Inorganic Chemistry, ETH Zurich*

PRODUCTION OF A THIN V-48 POSITRON SOURCE

Proton activation of a β^+ emitter foil for positron beams

L. Gerchow [1], P. Crivelli [1], A.M. Müller, M. Döbeli

In order to test a new technique to improve the conversion efficiency of a broad energy spectrum of beta decay into mono-energetic positrons a β^+ source on a thin foil (~1 µm thick) is required.

Fig. 1: *Special target holder designed for the foil irradiation. The W liner avoids activation of the holder and the large copper body takes up the thermal beam power.*

A careful evaluation showed that ^{48}V with a half-life of 16 days is suitable and can be produced via the reaction ^{48}Ti(p,n)^{48}V by irradiation of a titanium foil with protons at 8 MeV. The cross-section of this reaction is 295 mb. On the one hand this activation pathway produces only very little parasitic activity, on the other hand excessive activation of the LIP accelerator beam guiding system can be avoided at this very moderate proton energy. A special target holder was designed for the experiment. It has a solid copper body which can take up the thermal power of the beam and is lined with tungsten sheet metal to avoid activation by the beam transmitted through the foil or by stray particles. A second irradiation position allowed precise positioning of the beam spot with the mark produced by the short exposure of a piece of paper. The holder is slanted by 45° towards the incident beam direction to increase the effective target thickness during irradiation. A 1 µm thick Ti foil was irradiated in a spot of about 5 mm^2 with 5.2 mC of protons at 8 MeV. The activity measurement with a Ge detector (Fig. 2) revealed that only very short lived parasitic radionuclides were produced in the foil and the actual ^{48}V activity was 11.9 kBq in very good agreement with 12.5 kBq expected from the published reaction cross-section.

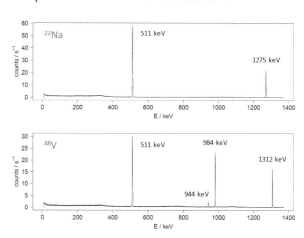

Fig. 2: *γ-spectrum of the activated foil compared to the spectrum of a ^{22}Na source of known activity. Very little parasitic activity is detected.*

The activation procedure turned out to be very effective. An improvement of almost an order of magnitude for the conversion efficiency of a broad energy spectrum of beta decay into mono-energetic positrons was achieved. Preliminary simulations show that another order of magnitude could be possible with an optimized setup, thus new sources on thin foils would be required in the near future for further testing. Such an improvement would be a breakthrough in the positron community allowing for the use of sources with much lower activities.

[1] *Institute of Particle Physics, ETH Zurich*

EDUCATION

Fruits, beetles and the Unteraargletscher

ETH Project: Studienwoche 2015

Medieval past in the heart of Zurich

FRUITS, BEETLES AND THE UNTERAARGLETSCHER

^{14}C ages of fragments of an early Holocene peat bog

N. Schürch[1], I. Hajdas, B. Schlüchter[1], C. Schlüchter[2], R. Drescher-Schneider [3]

Pieces of peat found in the glacier foreland of the Unteraargletscher in the Bernese Oberland were investigated in a research school project called Maturaarbeit. The fragments of peat originated from a bog that once existed in the vicinity of the glacier and was then later overrun by the advancing ice. Radiocarbon dating was employed to estimate the time of peat bog formation. The aim of this study was to investigate macro remains preserved in the peat. Fruits and insect remains provide a detailed picture of the surroundings of the glacier in the early Holocene.

Peat fragments were soaked in potassium hydroxide and sieved through the set of different sieves (4 mm, 2 mm and 0.5 mm) in order to separate the macro remains. The remnants, which remained on the sieves were investigated under microscope. Among collected macro remains were numerous fruits and remains of insects, sometimes even completely preserved (Fig. 1).

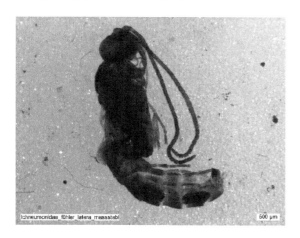

Fig. 1: *Completely preserved remains of a wasp Attractodes Mesoleptus found in peat fragment. (Photo by N. Schürch 2015).*

Various specialists were involved in identification of the fruit and insect remains (Ö.

Akeret; IPNA Basel, S. Klopfstein, H. Baur, Ch. Germann; nmbe Bern, V. Puthz; Naturkunde-museum Kassel, C. van Achterberg; NBC, Leiden, and R. Bryner; Biel). Nearly 650 fruits found belonging to the family of sedges were selected. Species determinations were not possible. The insect remains were to a large extent fragments of rove beetles (Staphylinidae). Amazingly, however, parasitic wasp (Ichneumonidae and Braconidae) (Fig 1 and 2, respectively), and a butterfly caterpillar were also found.

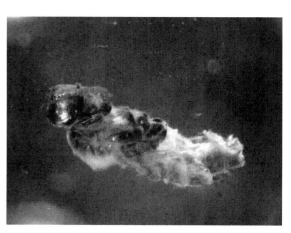

Fig. 2: *Remains of Braconidae wasp. A reflection of light on the eye is visible. Blurred parts is the thorax (Photo by N. Schürch 2015).*

Two samples of macrofossils provided ^{14}C ages of early Holocene. Based on these results it can be concluded that at this time the vegetation in the vicinity of the bog was dominated by sedges. An absence of macro remains from trees is striking. Also insects' fragments confirmed this landscape as all the species of beetle found preferred wetlands, swamps and marshes.

[1] *Standort Schadau, Gymnasium Thun*
[2] *Geology, University of Bern*
[3] *Plant Sciences, Karl-Franzens-Universität Graz*

ETH PROJECT: STUDIENWOCHE 2015

^{14}C content in 'very old' wood and 'young' leaves

I. Hajdas, N. Burri[1], L. Klarner[2], M. Maurer, M. Roldan, A. Synal, L. Wacker

The first week of June is traditionally dedicated to participants of the ETH project: *Studienwoche* for high school students who are interested in science. This year, experiments included sampling of various objects: a historic bag, fragment of compressed wood and fresh leaves (Fig. 1). These objects proved to be of a wide range radiocarbon ages.

Fig. 1: *Fresh leaves were collected at the ETH campus and prepared for ^{14}C analysis.*

Fig. 2: *All steps of preparation were carefully recorded.*

Independent of the age all the samples were carefully treated (Fig. 2) and prepared for the AMS analysis. Most importantly, the piece of compressed fossilized wood was treated to remove modern contamination. The standard treatment of ABA (acid-base-acid) was sufficient

to show that the piece found in Turbenthal is older than 50 thousand years. In fact concentrations of ^{14}C found in this wood were on the level of the so called blank material (^{14}C free).

On the other hand fresh leaves were treated with acid only to remove potential contamination through dust containing carbonate particles. The measured ^{14}C content shows that, due to fossil fuel combustion and old CO_2 added to the atmosphere, the atmospheric ^{14}C content is nearly at the level of pre-nuclear tests (Fig. 3).

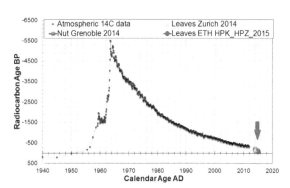

Fig. 3: *Present day atmosphere shows very low ^{14}C content, this is best illustrated by the measurements on fresh leaves growing in 2014 and 2015 (red arrow).*

---

[1] *Kantonsschule im Lee, Turbenthal*
[2] *Deutsche Schule Moskau, Russia*

MEDIEVAL PAST IN THE HEART OF ZURICH

Students excursion to the Münsterhof excavation site in Zurich

I. Hajdas, S. Ivy-Ochs

Zurich is a very dynamic town with a long history (Fig. 1). This fortunate mixture of past and present is visible in the development of the town. Whenever new urban concepts are being implemented the town's past comes to light.

Fig. 1: *Painting of Einsiedlerhof (by Hans Leu around 1500 AD), erected 1234–67 AD. (https://www.stadt-zuerich.ch).*

The 2015 reconstructions at the Münsterhof, which replaced the parking lots (Fig. 2) by a car free space and a fountain, were closely followed by the citizens of Zurich.

Fig. 2: *Parking lots in front of the Münsterhof 2004 (picture AfS/Archäologie).*

The unique opportunities of public visits provided by the Archaeological Services are of great cultural and educational value. The excavations are closely related to activities of the LIP laboratory as some of the samples of bone, wood and charcoal are subject to ^{14}C dating. A visit with students of the lecture *Quaternary Dating Methods* (ETH Zurich, fall term 2015) gave them a chance to get a direct insight into problems encountered in the field as well as a unique opportunity to see the sediments and medieval layers (Fig. 3) that soon will be covered for many decades to come.

Fig. 3: *Visit to the excavation site. A group of student participants on the guided tour of Jonathan Frey (Stadt-Zurich).*

PUBLICATIONS

M.L. Adamic, T.E. Lister, E.J. Dufek, D.D. Jenson, J.E. Olson, C. Vockenhuber and M.G. Watrous
Electrodeposition as an alternate method for preparation of environmental samples for iodide by AMS
Nuclear Instruments and Methods in Physics Research Section B: Beam Interactions with Materials and Atoms **361** (2015) 372-375

E. Bard, T. Tuna, Y. Fagault, L. Bonvalot, L. Wacker, S. Fahrni and H.-A. Synal
AixMICADAS, the accelerator mass spectrometer dedicated to ^{14}C recently installed in Aix-en-Provence, France
Nuclear Instruments and Methods in Physics Research Section B: Beam Interactions with Materials and Atoms **361** (2015) 80-86

M. Boudin, M. Van Strydonck, T. van den Brande, H.-A. Synal and L. Wacker
RICH–A new AMS facility at the Royal Institute for Cultural Heritage, Brussels, Belgium
Nuclear Instruments and Methods in Physics Research Section B: Beam Interactions with Materials and Atoms **361** (2015) 120-123

L. Calcagnile, G. Quarta, L. Maruccio, H.-A. Synal and A.M. Müller
The new AMS system at CEDAD for the analysis of ^{10}Be, ^{26}Al, ^{129}I and actinides: Set-up and performances
Nuclear Instruments and Methods in Physics Research Section B: Beam Interactions with Materials and Atoms **361** (2015) 100-104

E. Chamizo, M. Christl and L.K. Fifield
Measurement of ^{236}U on the 1 MV AMS system at the Centro Nacional de Aceleradores (CNA)
Nuclear Instruments and Methods in Physics Research Section B: Beam Interactions with Materials and Atoms **358** (2015) 45-51

J. Chen, Y. Lv, M. Döbeli, Y. Li, X. Shi and L. Chen
Composition control of pulsed laser deposited copper (I) chalcogenide thin films via plasma/Ar interactions
Science China-Materials **58** (2015) 263-268

J. Chen, D. Stender, M. Pichler, M. Döbeli, D. Pergolesi, C.W. Schneider, A. Wokaun and T. Lippert
Tracing the plasma interactions for pulsed reactive crossed-beam laser ablation
Journal of Applied Physics **118** (2015) 165306

J. Chen, Y. Zhou, S. Middey, J. Jiang, N. Chen, L. Chen, X. Shi, M. Döbeli, J. Shi, J. Chakhalian and S. Ramanathan
Self-limited kinetics of electron doping in correlated oxides
Applied Physics Letters **107** (2015) 031905

M. Christl, N. Casacuberta, J. Lachner, S. Maxeiner, C. Vockenhuber, H.-A. Synal, I. Goroncy, J. Herrmann, A. Daraoui, C. Walther and R. Michel
Status of ^{236}U analyses at ETH Zurich and the distribution of ^{236}U and ^{129}I in the North Sea in 2009
Nuclear Instruments and Methods in Physics Research Section B: Beam Interactions with Materials and Atoms **361** (2015) 510-516

M. Christl, N. Casacuberta, C. Vockenhuber, C. Elsässer, P. Bailly du Bois, J. Herrmann and H.-A. Synal
Reconstruction of the ^{236}U input function for the Northeast Atlantic Ocean: Implications for $^{129}I/^{236}U$ and $^{236}U/^{238}U$-based tracer ages
Journal of Geophysical Research: Oceans **120** (2015) 7282-7299

A. Ciner, U. Dogan, C. Yildirim, N. Akcar, S. Ivy-Ochs, V. Alfimov, P.W. Kubik and C. Schluechter
Quaternary uplift rates of the Central Anatolian Plateau, Turkey: insights from cosmogenic isochron-burial nuclide dating of the Kizilirmak River terraces
Quaternary Science Reviews **107** (2015) 81-97

R. Cusnir, P. Steinmann, M. Christl, F. Bochud and P. Froidevaux
Speciation and Bioavailability Measurements of Environmental Plutonium Using Diffusion in Thin Films
Journal of visualized experiments: JoVE (2015)

X. Dai, M. Christl, S. Kramer-Tremblay and H.-A. Synal
Ultra-trace determination of neptunium-237 and plutonium isotopes in urine samples by compact accelerator mass spectrometry
CNL Nuclear Review **4** (2015) 125-130

V. Dausmann, M. Frank, C. Siebert, M. Christl and J.R. Hein
The evolution of climatically driven weathering inputs into the western Arctic Ocean since the late Miocene: Radiogenic isotope evidence
Earth and Planetary Science Letters **419** (2015) 111-124

P. Deline, N. Akcar, S. Ivy-Ochs and P.W. Kubik
Repeated Holocene rock avalanches onto the Brenva Glacier, Mont Blanc massif, Italy: A chronology
Quaternary Science Reviews **126** (2015) 186-200

B.A. Dittmann, T.J. Dunai, A. Dewald, S. Heinze, C. Feuerstein, E. Strub, L.K. Fifield, M.B. Froehlich, S.G. Tims, A. Wallner and M. Christl
Preparation of a multi-isotope plutonium AMS standard and preliminary results of a first inter-lab comparison
Nuclear Instruments and Methods in Physics Research Section B: Beam Interactions with Materials and Atoms **361** (2015) 327-331

C. Elsässer, D. Wagenbach, I. Levin, A. Stanzick, M. Christl, A. Wallner, S. Kipfstuhl, I. Seierstad, H. Wershofen and J. Dibb
Simulating ice core ^{10}Be on the glacial–interglacial timescale
Climate of the Past **11** (2015) 115-133

D. Fu, X. Ding, K. Liu, A.M. Müller, M. Suter, M. Christl, L. Zhou and H.-A. Synal
Further improvement for ^{10}Be measurement on an upgraded compact AMS radiocarbon facility
Nuclear Instruments and Methods in Physics Research Section B: Beam Interactions with Materials and Atoms (2015)

R.-H. Fülöp, L. Wacker and T. Dunai
Progress report on a novel in situ ^{14}C extraction scheme at the University of Cologne
Nuclear Instruments and Methods in Physics Research Section B: Beam Interactions with Materials and Atoms **361** (2015) 20-24

E.F. Gjermundsen, J.P. Briner, N. Akcar, J. Foros, P.W. Kubik, O. Salvigsen and A. Hormes
Minimal erosion of Arctic alpine topography during late Quaternary glaciation
Nature Geoscience **8** (2015) 789-792

A. Graf, N. Akcar, S. Ivy-Ochs, S. Strasky, P.W. Kubik, M. Christl, M. Burkhard, R. Wieler and C. Schluechter
Multiple advances of Alpine glaciers into the Jura Mountains in the Northwestern Switzerland
Swiss Journal of Geosciences **108** (2015) 225-238

D. Güttler, F. Adolphi, J. Beer, N. Bleicher, G. Boswijk, M. Christl, A. Hogg, J. Palmer, C. Vockenhuber and L. Wacker
Rapid increase in cosmogenic ^{14}C in AD 775 measured in New Zealand kauri trees indicates short-lived increase in ^{14}C production spanning both hemispheres
Earth and Planetary Science Letters **411** (2015) 290-297

N. Haghipour, J.-P. Burg, S. Ivy-Ochs, I. Hajda, P.W. Kubik and M. Christl
Correlation of fluvial terraces and temporal steady-state incision on the onshore Makran accretionary wedge in southeastern Iran: Insight from channel profiles and ^{10}Be exposure dating of strath terraces
GSA Bulletin **127** (2015) 560-583

B. Hammer-Rotzler, J. Neuhausen, C. Vockenhuber, V. Boutellier, M. Wohlmuther, A. Türler and D. Schumann
Radiochemical determination of ^{129}I and ^{36}Cl in MEGAPIE, a proton irradiated lead-bismuth eutectic spallation target
Radiochimica Acta **103** (2015) 745-758

K. Hippe, A. Möller, A. von Quadt, I. Peytcheva and K. Hammerschmidt
Zircon geochronology of the Koraput alkaline complex: Insights from combined geochemical and U–Pb–Hf isotope analyses, and implications for the timing of alkaline magmatism in the Eastern Ghats Belt, India
Gondwana Research (2015)

J. Hwang, M. Kim, S.J. Manganini, C.P. McIntyre, N. Haghipour, J. Park, R.A. Krishfield, R.W. Macdonald, F.A. McLaughlin and T.I. Eglinton
Temporal and spatial variability of particle transport in the deep Arctic Canada Basin
Journal of Geophysical Research: Oceans **120** (2015) 2784-2799

S. Ivy-Ochs
Glacier variations in the European Alps at the end of the last glaciation
Cuadernos de investigación geográfica **41** (2015) 295-315

T. Jäger, Y.E. Romanyuk, S. Nishiwaki, B. Bissig, F. Pianezzi, P. Fuchs, C. Gretener, M. Döbeli and A.N. Tiwari
Hydrogenated indium oxide window layers for high-efficiency Cu(In,Ga)Se2 solar cells
Journal of Applied Physics **117** (2015) 205301

M. Jiskra, J.G. Wiederhold, U. Skyllberg, R.-M. Kronberg, I. Hajdas and R. Kretzschmar
Mercury Deposition and Re-emission Pathways in Boreal Forest Soils Investigated with Hg Isotope Signatures
Environmental Science & Technology **49** (2015) 7188-7196

R.S. Jones, A.N. Mackintosh, K.P. Norton, N.R. Golledge, C.J. Fogwill, P.W. Kubik, M. Christl and S.L. Greenwood
Rapid Holocene thinning of an East Antarctic outlet glacier driven by marine ice sheet instability
Nature Communications **6** (2015)

F. Kober, G. Zeilinger, K. Hippe, O. Marc, T. Lendzioch, R. Grischott, M. Christl, P.W. Kubik and R. Zola
Tectonic and lithological controls on denudation rates in the central Bolivian Andes
Tectonophysics **657** (2015) 230-244

A. Kounov, S. Niedermann, M.J. De Wit, A.T. Codilean, G. Viola, M. Andreoli and M. Christl
Cosmogenic ^{21}Ne and ^{10}Be reveal a more than 2 Ma alluvial fan flanking the Cape Mountains, South Africa
South African Journal of Geology **118** (2015) 129-144

K. Krainer, D. Bressan, B. Dietre, J.N. Haas, I. Hajdas, K. Lang, V. Mair, U. Nickus, D. Reidl, H. Thies and D. Tonidandel
A 10,300-year-old permafrost core from the active rock glacier Lazaun, southern Otztal Alps (South Tyrol, northern Italy)
Quaternary Research **83** (2015) 324-335

J. Leifeld, M. Heiling and I. Hajdas
Age and thermal stability of particular organic matter fractions indicate the presence of black carbon in soil
Radiocarbon **57** (2015) 99-107

M. Lupker, K. Hippe, L. Wacker, F. Kober, C. Maden, R. Braucher, D. Bourlès, J.V. Romani and R. Wieler
Depth-dependence of the production rate of in situ ^{14}C in quartz from the Leymon High core, Spain
Quaternary Geochronology **28** (2015) 80-87

S.K. Mandal, M. Lupker, J.-P. Burg, P.G. Valla, N. Haghipour and M. Christl
Spatial variability of 10Be-derived erosion rates across the southern Peninsular Indian escarpment: A key to landscape evolution across passive margins
Earth and Planetary Science Letters **425** (2015) 154-167

P.J. Mann, T.I. Eglinton, C.P. McIntyre, N. Zimov, A. Davydova, J.E. Vonk, R.M. Holmes and R.G. Spencer
Utilization of ancient permafrost carbon in headwaters of Arctic fluvial networks
Nature communications **6** (2015)

S. Maxeiner, M. Seiler, M. Suter and H.-A. Synal
Charge state distributions and charge exchange cross sections of carbon in helium at 30–258 keV
Nuclear Instruments and Methods in Physics Research Section B: Beam Interactions with Materials and Atoms **361** (2015) 541-547

S. Maxeiner, M. Suter, M. Christl and H.-A. Synal
Simulation of ion beam scattering in a gas stripper
Nuclear Instruments and Methods in Physics Research Section B: Beam Interactions with Materials and Atoms **361** (2015) 237-244

F. Mekhaldi, R. Muscheler, F. Adolphi, A. Aldahan, J. Beer, J.R. McConnell, G. Possnert, M. Sigl, A. Svensson, H.-A. Synal, K.C. Welten and T.E. Woodruff
Multiradionuclide evidence for the solar origin of the cosmic-ray events of AD 774/5 and 993/4
NATURE COMMUNICATIONS **DOI: 10.1038/ncomms9611** (2015) 1-8

R. Michel, A. Daraoui, M. Gorny, D. Jakob, R. Sachse, L. Romantschuk, V. Alfimov and H.-A. Synal
Retrospective dosimetry of Iodine-131 exposures using Iodine-129 and Caesium-137 inventories in soils–A critical evaluation of the consequences of the Chernobyl accident in parts of Northern Ukraine
Journal of environmental radioactivity **150** (2015) 20-35

A.M. Müller, M. Christl, J. Lachner, H.-A. Synal, C. Vockenhuber and C. Zanella
^{26}Al measurements below 500 kV in charge state 2+
Nuclear Instruments and Methods in Physics Research Section B: Beam Interactions with Materials and Atoms **361** (2015) 257-262

A.M. Müller, M. Döbeli, M. Seiler and H.-A. Synal
A simple Bragg detector design for AMS and IBA applications
Nuclear Instruments and Methods in Physics Research Section B: Beam Interactions with Materials and Atoms **356–357** (2015) 81-87

J. Nagelisen, J.R. Moore, C. Vockenhuber and S. Ivy-Ochs
Post-glacial rock avalanches in the Obersee Valley, Glarner Alps, Switzerland
Geomorphology **238** (2015) 94-111

F.C. Nunes, R. Delunel, F. Schlunegger, N. Akcar and P.W. Kubik
Bedrock bedding, landsliding and erosional budgets in the Central European Alps
Terra Nova **27** (2015) 370-378

A. Ojeda-GP, C.W. Schneider, M. Döbeli, T. Lippert and A. Wokaun
The flip-over effect in pulsed laser deposition: Is it relevant at high background gas pressures?
Applied Surface Science **357** (2015) 2055-2062

A. Ojeda-G-P, C.W. Schneider, M. Döbeli, T. Lippert and A. Wokaun
Angular distribution of species in pulsed laser deposition of LaxCa1–xMnO3
Applied Surface Science **336** (2015) 150-156

R. Pellitero, B.R. Rea, M. Spagnolo, J. Bakke, S. Ivy-Ochs, P. Hughes, S. Lukas and A. Ribolini
A GIS tool for automatic calculation of glacier equilibrium-line altitudes
Computers & Geosciences (2015)

P. Reinhard, B. Bissig, F. Pianezzi, E. Avancini, H. Hagendorfer, D. Keller, P. Fuchs, M. Döbeli, C. Vigo, P. Crivelli, S. Nishiwaki, S. Buecheler and A. Tiwari
Features of KF and NaF Postdeposition Treatments of Cu (In, Ga) Se2 Absorbers for High Efficiency Thin Film Solar Cells
Chemistry of Materials **27** (2015) 5755-5764

M. Schaller, J. Lachner, M. Christl, C. Maden, N. Spassov, A. Ilg and M. Böhme
Authigenic Be as a Tool to Date River Terrace Sediments?–An Example From a Late Miocene Hominid Locality in Bulgaria
Quaternary Geochronology (2015)

B. Scherrer, M. Döbeli, P. Felfer, R. Spolenak, J. Cairney and H. Galinski
The hidden pathways in dense energy materials – Oxygen at defects in nanocrystalline metals
Advanced Materials **27** (2015) 6220-6224

A. Schild, I. Herter-Aeberli, K. Fattinger, S. Anderegg, T. Schulze-König, C. Vockenhuber, H.-A. Synal,
H. Bischoff-Ferrari, P. Weber and A. von Eckardstein
Oral Vitamin D Supplements Increase Serum 25-Hydroxyvitamin D in Postmenopausal Women and Reduce Bone Calcium Flux Measured by ^{41}Ca Skeletal Labeling
The Journal of Nutrition **145** (2015) 2333-2340

M. Seiler, S. Maxeiner, L. Wacker and H.-A. Synal
Status of mass spectrometric radiocarbon detection at ETHZ
Nuclear Instruments and Methods in Physics Research Section B: Beam Interactions with Materials and Atoms **361** (2015) 245-249

C. Solís, E. Chávez, M. Ortiz, E. Andrade, E. Ortíz, S. Szidat and L. Wacker
AMS-^{14}C analysis of graphite obtained with an Automated Graphitization Equipment (AGE III) from aerosol collected on quartz filters
Nuclear Instruments and Methods in Physics Research Section B: Beam Interactions with Materials and Atoms **361** (2015) 419-422

O.N. Solomina, R.S. Bradley, D.A. Hodgson, S. Ivy-Ochs, V. Jomelli, A.N. Mackintosh, A. Nesje, L.A. Owen,
H. Wanner, G.C. Wiles and N.E. Young
Holocene glacier fluctuations
Quaternary Science Reviews **111** (2015) 9-34

R.G. Spencer, P.J. Mann, T. Dittmar, T.I. Eglinton, C. McIntyre, R.M. Holmes, N. Zimov and A. Stubbins
Detecting the signature of permafrost thaw in Arctic rivers
Geophysical Research Letters **42** (2015) 2830-2835

C. Stalder, A. Vertino, A. Rosso, A. Rüggeberg, C. Pirkenseer, J.E. Spangenberg, S. Spezzaferri, O. Camozzi,
S. Rappo and I. Hajdas
Microfossils, a Key to Unravel Cold-Water Carbonate Mound Evolution through Time: Evidence from the Eastern Alboran Sea
PloS one **10** (2015) e0140223

G. Steinhauser, T. Niisoe, K.H. Harada, K. Shozugawa, S. Schneider, H.-A. Synal, C. Walther, M. Christl,
K. Nanba, H. Ishikawa and A. Koizumi
Post-accident sporadic releases of airborne radionuclides from the Fukushima Daiichi Nuclear Power Plant Site
Environmental science & technology **49** (2015) 14028-14035

E. Strub, H. Wiesel, G. Delisle, S.A. Binnie, A. Liermann, T.J. Dunai, U. Herpers, A. Dewald, S. Heinze,
M. Christl and H.H. Coenen
Glaciation history of Queen Maud Land (Antarctica) – New exposure data from nunataks
Nuclear Instruments and Methods in Physics Research Section B: Beam Interactions with Materials and Atoms **361** (2015) 599-603

S. Tao, T.I. Eglinton, D.B. Montluçon, C. McIntyre and M. Zhao
Pre-aged soil organic carbon as a major component of the Yellow River suspended load: Regional significance and global relevance
Earth and Planetary Science Letters **414** (2015) 77-86

V. Vanacker, F. von Blanckenburg, G. Govers, A. Molina, B. Campforts and P.W. Kubik
Transient river response, captured by channel steepness and its concavity
Geomorphology **228** (2015) 234-243

C. Vockenhuber, N. Casacuberta, M. Christl and H.-A. Synal
Accelerator Mass Spectrometry of ^{129}I towards its lower limits
Nuclear Instruments and Methods in Physics Research Section B: Beam Interactions with Materials and Atoms **361** (2015) 445-449

C. Vockenhuber, T. Schulze-König, H.-A. Synal, I. Aeberli and M.B. Zimmermann
Efficient ^{41}Ca measurements for biomedical applications
Nuclear Instruments and Methods in Physics Research Section B: Beam Interactions with Materials and Atoms **361** (2015) 273-276

J.E. Vonk, L. Giosan, J. Blusztajn, D. Montlucon, E.G. Pannatier, C. McIntyre, L. Wacker, R.W. Macdonald, M.B. Yunker and T.I. Eglinton
Spatial variations in geochemical characteristics of the modern Mackenzie Delta sedimentary system
Geochimica et Cosmochimica Acta **171** (2015) 100-120

H. Wittmann, F. Blanckenburg, N. Dannhaus, J. Bouchez, J. Gaillardet, J. Guyot, L. Maurice, H. Roig, N. Filizola and M. Christl
A test of the cosmogenic ^{10}Be (meteoric)/^9Be proxy for simultaneously determining basin-wide erosion rates, denudation rates, and the degree of weathering in the Amazon basin
Journal of Geophysical Research: Earth Surface (2015)

J. Zalasiewicz, C.N. Waters, M. Williams, A.D. Barnosky, A. Cearreta, P. Crutzen, E. Ellis, M.A. Ellis, I.J. Fairchild, J. Grinevald, P.K. Haff, I. Hajdas, R. Leinfelder, J. McNeill, E.O. Odada, C. Poirier, D. Richter, W. Steffen, C. Summerhayes, J.P.M. Syvitski, D. Vidas, M. Wagreich, S.L. Wing, A.P. Wolfe, A. Zhisheng and N. Oreskes
When did the Anthropocene begin? A mid-twentieth century boundary level is stratigraphically optimal
Quaternary International (2015)

B. Zollinger, C. Alewell, C. Kneisel, K. Meusburger, D. Brandová, P.W. Kubik, M. Schaller, M. Ketterer and M. Egli
The effect of permafrost on time-split soil erosion using radionuclides (^{137}Cs, $^{239+240}Pu$, meteoric ^{10}Be) and stable isotopes ($\delta^{13}C$) in the eastern Swiss Alps
Journal of Soils and Sediments (2015) 1-20

TALKS AND POSTERS

N. Akçar, S. Ivy-Ochs, V. Alfimov, A. Claude, R. Reber, M. Christl, C. Vockenhuber, F. Schlunegger, R. Meinert, A. Dehnert, C. Schlüchter
Cosmogenic nuclide dating of glaciofluvial deposits: insights from the Alps
Austria, Vienna, 12.-17.04.2015, EGU General Assembly

N. Akçar, O. Fredin, A. Romundset, R. Reber, S. Ivy-Ochs, M. Christl, F. Schlunegger, C. Schlüchter
Early break-up of the Norwegian Channel Ice Stream and deglaciation of southernmost Norway: Insights from cosmogenic depth-profile dating
Japan, Nagoya, 26.07.2015 – 02.08.2015, XIX. Inqua Congress

N. Akçar, S. Ivy-Ochs, V. Alfimov, R. Reber, A. Claude, M. Christl, C. Vockenhuber, F. Schlunegger, R. Meinert, A. Dehnert, C. Schlüchter
Isochron-burial dating of glaciofluvial deposits: first results from the Alps
Japan, Nagoya, 26.07.2015 – 02.08.2015, XIX. Inqua Congress

N. Akçar, N. Mozafari Amiri, S. Ivy-Ochs, C. Vockenhuber, D. Tikhomirov, Ç. Özkaymak, Ö. Sümer, B. Uzel, H. Sözbilir
Revealing the seismically active periods beyond the historical archives with cosmogenic 36Cl
Norway, Trondheim, 24.04.2015, Goldschmidt Lectures - NGU

N. Akçar, V. Yavuz, S. Ivy-Ochs, F. Nyffenegger, O. Fredin, F. Schlunegger
Rearward landsliding in sensitive clays: February 2011 massive failures at the Çöllolar coalfield, eastern Turkey
Switzerland, Basel, 20.-21.11.2015, Swiss Geoscience Meeting

S. Aksay, S. Ivy-Ochs, K. Hippe, L. Grämiger, C. Vockenhuber
The geomorphological evolution of a landscape in a tectonically active region: The Sennwald Landslide
Switzerland, Basel, 20.-21.11.2015, Swiss Geoscience Meeting

C. Bayrakdar, N. Akçar, T. Görüm, S Ivy-Ochs, C. Vockenhuber
Glacio-Karstic and chronological evolution of the Akdağ rockslide (SW Turkey)
Austria, Vienna, 12.-17.04.2015, EGU General Assembly

C. Bayrakdar, N. Akçar, T. Görüm, S. Ivy-Ochs, C. Vockenhuber
Morphology and chronology of the Akdag Landslide (SW Turkey)
Japan, Nagoya, 26.07.2015 – 02.08.2015, XIX. Inqua Congress

M.G. Bichler, M. Reindl, J.M. Reitner, S. Ivy-Ochs
Defining the Lateglacial stratigraphy in the Eastern-Alps using gravitational and glacial sedimentation sequences
Austria, Vienna, 12.-17.04.2015, EGU General Assembly

M. Boxleitner, M. Maisch, P. Walthard, S. Ivy-Ochs, D. Brandova, M. Egli
Lateglacial and Holocene glacier development and landscape evolution in Meiental, Uri (CH)
Switzerland, Basel, 20.-21.11.2015, Swiss Geoscience Meeting

N. Casacuberta, M. R. Van-der-Loeff, P. Masqué, J. Herrmann, J. Lachner, G. Henderson, C. Walther, C. Vockenhuber, H.-A. Synal, M. Christl
^{236}U as a new oceanographic tracer: First data in the North Sea, the Arctic Ocean and the Atlantic Ocean
Spain, Granada, 20.02.2015, 2015 ASLO Aquatic Science Meeting

M. Castrillejo, N. Casacuberta, M. Christl, C. Vockenhuber, H.-A. Synal, P. Masqué J. Garcia-Orellana
First Comprehensive mapping of U-236 and I-129 in the Mediterranean Sea
Spain, Granada, 20.02.2015, 2015 ASLO Aquatic Science Meeting

M. Castrillejo, N. Casacuberta, M. Christl, J. Garcia-Orellana, P. Masqué, C. Vockenhuber, H.-A. Synal,
First Transect of ^{236}U and ^{129}I in the Mediterranean Sea.
Czech Republic, Prague, 18.08.2015, Goldschmidt Conference

N. Casacuberta, M. Christl, J. Garcia-Orellana, P. Masqué, C. Vockenhuber, H.-A. Synal,
First transect of ^{236}U and ^{129}I in the Mediterranean Sea
Czech Republic, Prague, 18.08.2015, Goldschmidt Conference

N. Casacuberta, J. Lachner, S. Maxeiner, C. Vockenhuber, J. Herrmann, M. Castrillejo, P. Masque, M. Rutgers van der Loeff, H.-A. Synal
Die Verteilung von ^{236}U/^{238}U im Nordatlantik und den angrenzenden Ozeanen
Germany, Heidelberg, 25.03.2015, DPG Spring Meeting

N. Casacuberta, C. Vockenhuber, P. Masque, M. Rutgers van der Loeff, G.M. Henderson, H.-A. Synal
The distribution of ^{236}U/^{238}U and ^{129}I/^{236}U in the North Atlantic and Arctic Oceans
Czech Republic, Prague, 18.08.2015, Goldschmidt Conference

A. Çiner, U. Doğan, C. Yıldırım, N. Akçar, S. Ivy-Ochs, V. Alfimov, P.W. Kubik, C. Schlüchter
Kızılırmak nehir şekillerinden elde edilen kozmojenik izokron gömülme yaşları ışığında Orta Anadolu Platosunun Kuvaterner yükselim hız oranları
Turkey, Ankara, 06.-10.04.2015, 68th Geological Congress of Turkey

A. Çiner, U. Doğan, C. Yıldırım, N. Akçar, S. Ivy-Ochs, V. Alfimov, P.W. Kubik, C. Schlüchter
Quaternary uplift rates of the Central Anatolian Plateau, Turkey: Insights from cosmogenic isochron-burial nuclide dating of the Kizilirmak River terraces
Japan, Nagoya, 26.07.2015 – 02.08.2015, XIX. Inqua Congress

C. Claude, N. Akçar, S. Ivy-Ochs, F. Schlunegger, P.W. Kubik , M. Christl, M. Rahn, A. Dehnert, C. Schlüchter
Depth-profile dating of proximal glaciofluvial gravels in the northern Swiss Alpine Foreland
Japan, Nagoya, 26.07.2015 – 02.08.2015, XIX. Inqua Congress

C. Claude, N. Akçar, S. Ivy-Ochs, F. Schlunegger, M. Rahn, Dehnert A., C. Schlüchter
Quaternary landscape evolution of the Swiss Alpine Foreland
Japan, Nagoya, 26.07.2015 – 02.08.2015, XIX. Inqua Congress

C. Claude, N. Akçar, S. Ivy-Ochs, F. Schlunegger, P.W. Kubik, M. Christl, C. Vockenhuber, M. Rahn, A. Dehnert, C. Schlüchter
Long-term bedrock incision rates in the northern Swiss Alpine Foreland inferred from reconstructed Deckenschotter chronologies
Switzerland, Basel, 20.-21.11.2015, Swiss Geoscience Meeting

B. Courel, P. Adam, Ph. Schaeffer, C. Féliu, S. M. Bernasconi, I. Hajdas
Investigation of the content of protohistoric silos from the Bronze and Iron Age in Alsace (NE France): a biomarker approach
Czech Republic, Prague, 13.-18.09.2015, 27th IMOG Meeting

A. Daraoui, C. Walther, C. Vockenhuber, H.-A. Synal
Iodine-129 in the atmosphere on the Zugspitze
Germany, Heidelberg, 27.03.2015, DPG Spring Meeting

A. Daraoui, B. Riebe, M. Gorny, C. Walther, K. Hürkamp, J. Tschiersch, C. Vockenhuber, H.-A. Synal
Vergleichbare ^{129}I-Konzentrationen und ^{129}I/^{127}I-Isotopenverhältnisse in Schnee von der Zugspitze und in Nordseewasser?
Germany, Dresden, 31.08.2015, GDCh-Tagung

R. Delunel, J. Casagrande, F. Schlunegger, N. Akçar, P.W. Kubik
Initiation age and incision rates of inner gorges: Do they record multiple glacial-interglacial cycles?
Austria, Vienna, 12.-17.04.2015, EGU General Assembly

B. Dietre, M. Hirnsperger, D. Bressan, Ch. Walser, I. Hajdas, K. Lang, V. Mair, U. Nickus, D. Reidl, H. Thies, D. Tonidandel, K. Krainer, J.-N. Haas
Palynology and Palaeoecology of the Holocene Rock Glacier at Lazaun (South Tyrol, Italy)
Japan, Nagoya, 26.07.2015 – 02.08.2015, XIX. Inqua Congress

M. Döbeli
Materials characterization by IBA methods
Germany, Dresden, 07.09.2015, EU-Project Sprite Course

L. Hendriks, I. Hajdas, M. Küffner, C. McIntyre, N. C Scherrer, E. S.B. Ferreira
Microscale radiocarbon dating of paintings
USA, Chicago, 21.-22.05.2015, MaSC 2015 Meeting

C. Fogwill, C. Turney, N. Golledge, D. Rood, K. Hippe, L. Wacker, R. Wieler, E. Rainsley, R. Jones
Mechanisms Driving abrupt shifts inWest Antarctic ice stream direction during the Holocene
Austria, Vienna, 16.04.2015, EGU General Assembly

F.A. Lechleitner, R.A. Jamieson, C.McIntyre, L.M. Baldini, U.L. Baldini, T.I.Eglinton
Modelling of dead carbon fraction in speleothems: a step towards reliable speleothem ^{14}C-chronologies
Austria, Vienna, 13.04.2015, EGU General Assembly

O. Fredin, N. Akçar, A. Romundset, R. Reber, S. Ivy-Ochs, P.W. Kubik, F. Høgaas, C. Schlüchter
A more complex deglaciation chronology of Southern Norway than previously thought. New geochronological constraints based on cosmogenic exposure ages of marginal moraines
Austria, Vienna, 12.-17.04.2015, EGU General Assembly

S. Hou, C. Wang, M. Christl, S. Maxeiner, H.-A. Synal, C. Vockenhuber
Chronology of ^{239}/^{240}Pu and of ^{236}U in the Miaoergou glacier from eastern Tien Shan, China
USA, Denver, 22.03.2015, ACS Meeting

M. George, M. Döbeli, A.M. Müller, M. Schulte-Borchers, H.-A. Synal
Digital Pulse Processing for Ion Beam Analysis at ETH Zurich
Croatia, Opatija, 15.06.2015, IBA Conference

R. Grischott, F. Kober, K. Hippe, M. Lupker, S. Ivy-Ochs, I. Hajdas, M. Christl
Climate control on alpine denudation in the Holocene – Clues from two converse datasets of paleo-CWDR
Switzerland, Basel, 21.11.2015, Swiss Geoscience Meeting

K. Gückel, T. Shinonaga, K. Hain, G. Korschinek, M. Christl
Determination of Plutonium and Americium in snow and rain
USA, Santa Fe, 13.-18.09.2015, Migration 2015 Conference

N. Haghipour, T. Eglinton, C. McIntyre,J. Darvishi Khatoono, D. Hunziker, A. Mohammadi
Austria, Vienna, 30.11.2015, EGU General Assembly

I. Hajdas, L. Hendriks , A. Fontana, G. Monegato
Evaluation of preparation methods in radiocarbon dating of old wood
Austria, Vienna, 12.-17.04.2016, EGU General Assembly

I. Hajdas, M. Maurer, M. Roldan Torres de Roettig
Review of recent treatment methods in AMS radiocarbon dating
Japan, Nagoya, 26.07.2015 – 02.08.2015, XIX. Inqua Congress

I. Hajdas, L. Hendriks , A. Fontana, G. Monegato
Evaluation of preparation methods in radiocarbon dating of old wood
Senegal, Dakar, 16.-20.11.2015, Radiocarbon Conference

I. Hajdas, M. Maurer, M. Roldan Torres de Roettig
Prospects for mortar ^{14}C dating at ETH Zurich
Switzerland, Zurich, 09.-11.09.2015, 4th International Workshop on Mortar Dating

I. Hajdas
Review of recent treatment methods in AMS 14Cdating —prospects for dating MUP transition.
France, Germolles, 24.-25.08.2015, Réunion 2015 PCR Paléolithique en Bourgogne méridionale

I. Hajdas
Reservoir Ages and Calibration: Treatment issues--Study cases and potential records for database
France, Paris, 01.07.2015, IntCal Focus Group on Marine Archives and Reservoir Ages

K. Hippe, K. Hammerschmidt, A. Möller, A. von Quadt, I. Peytcheva
Metamorphic zircon from the Indian Eastern Ghats Belt
Switzerland, Zurich, 10.05.2015, IGP Seminar (ETH ERDW)

K. Hippe, A. Fontana, I. Hajdas, S. Ivy Ochs
Reconstructing Alpine glacier activity during 50-20 ka BP by high-resolution radiocarbon dating of the Cormor alluvial megafan (Tagliamento glacier, NE Italy)
Switzerland, Basel, 21.11.2015, Swiss Geoscience Meeting

E. Huysecom, I. Hajdas, M.-A. Renold, A. Mayor, H.-A. Synal
The "enhancement" of cultural heritage by AMS dating: Ethical questions and practical proposals
Senegal, Dakar, 16.-20.11.2015, Radiocarbon Conference

S. Ivy-Ochs, S. Martin, P. Campedel, A. Viganò, S. Alberti, M. Rigo, C. Vockenhuber
The Marocche rock avalanches (Trentino, Italy)
Austria, Vienna, 12.-17.04.2015, EGU General Assembly

S. Ivy-Ochs, S. Martin, P. Campedel, A. Viganò, S. Alberti, M. Rigo, C. Vockenhuber
Age and geomorphology of the Marocche rock avalanches (Trentino, Italy)
Switzerland, Basel, 20.-21.11.2015, Swiss Geoscience Meeting

S. Ivy-Ochs
Surface exposure dating at the Flims and Tamins landslides, Switzerland
Switzerland, Flims, 04.-05.06.2015, Flims Landslide Workshop

S. Ivy-Ochs
What can we learn from bedrock? Combining cosmogenic ^{10}Be and ^{36}Cl
Switzerland, Lausanne, 04.11.2015, UniL Earth Surface Dynamics Group Seminar Series

S. Ivy-Ochs
Evaluating temporal trends in the timing of Younger Dryas glacier expansions across Europe
Switzerland, Riederalp, 30.09.2015, Leverhulme Network Meeting: Younger Dryas in Europe

F. Kober, K. Hippe, M. Christl, L. Wacker, W. Winkler, R. Lampe
Evaluating the in-situ produced cosmogenic nuclide inventory of longshore transported sand, Fischland-Darss-Zingst peninsula, southern Baltic Sea
Germany, Berlin, 05.10.2015, GeoBerlin

F. Kober, K. Hippe, B. Salcher, R. Grischott, S. Ivy-Ochs, S., R. Zurfluh, I. Hajdas, L. Wacker, M. Christl, N. Hählen
Sediment fingerprinting, mixing and geomorphic connectivity in alpine debris flow catchments (the Rotluai and Spreitlaui torrents; Guttannen)
Switzerland, Innertkirchen, 18.06.2015, Jahrestagung der SGmG 2015

O. Kronig, S. Ivy-Ochs, I.Hajdas, M. Christl, C. Schlüchter
Late Holocene evolution of the Triftjegletscher constrained with ^{10}Be exposure and radiocarbon dating
Switzerland, Basel, 21.11.2015, Swiss Geoscience Meeting

M. Gutjahr, P. Blaser, B. Antz, E. Böhm, M.L. de Carvalho Ferreira, F. Wombacher, M. Christl, S. Mulitza, S. Jaccard
Reconstructing past Ocean Circulation with ^{231}Pa/^{230}Th and Neodymium isotopes
Czech Republic, Prague, 18.08.2015, Goldschmidt Conference

M. Luetscher, S. Ivy-Ochs, M. Hof
Reconstructing the last deglaciation at Sieben Hengste, Switzerland
Switzerland, Basel, 20.-21.11.2015, Swiss Geoscience Meeting

S. Maxeiner, H.-A. Synal, M. Christl, M. Suter, A.M. Müller, C. Vockenhuber
Development of a multi isotope low energy AMS system
Germany, Heidelberg, 25.03.2015, DPG Spring Meeting

C. McIntyre, S. Fahrni, N. Haghipour, L. Wacker, M.Usman, T. Eglinton, H.-A. Synal
High-throughput, concurrent ^{14}C and ^{13}C analysis by EA-irMS/AMS: Earth science applications
Senegal, Dakar, 30.11.2015, Radiocarbon Conference

C. McIntyre, S. Fahrni, N. Haghipour, L. Wacker, M.Usman, T. Eglinton, and H.-A. Synal
High-throughput, concurrent ^{14}C and ^{13}C analysis by EA-irMS/AMS: Earth science applications
Senegal, Dakar, 30.11.2015, Radiocarbon Conference

M.S. Schwab, J.D. Rickli, J. Blusztajn, S. Manganini, H.R. Harvey, A. Forest, R.W. Macdonald, D. Vance, C. McIntyre, T.I. Eglinton
Coupled Organic & Inorganic Tracers of Particle Flux Processes in the Western Arctic Ocean
Czech Republic, Prague, 16.08.2015, Goldschmidt Conference

M. Molnar, I. Major, R. Janovics, K. Hubay, L. Rinyu, M. Veres, M. Seiler, L. Wacker, and A.J.T. Jull
Microsample C14 AMS analyses using gas ion source at HEKAL Laboratory
Senegal, Dakar, 30.11.2015, Radiocarbon Conference

N. Mozafari Amiri, D. Tikhomirov, Ç. Özkaymak, Ö. Sümer, B. Uzel, S. Ivy-Ochs, C. Vockenhuber, H. Sözbilir, N. Akçar
Revealing the seismically active periods beyond the historical archives: Fault scarp dating with 36Cl
Japan, Nagoya, 26.07.2015 – 02.08.2015, XIX. Inqua Congress

N. Mozafari Amiri, Ö. Sümer, D. Tikhomirov, Ç. Özkaymak, B. Uzel, S. Ivy-Ochs, C. Vockenhuber, H. Sözbilir, N. Akçar
Determination of paleoseismic activity with cosmogenic ^{36}Cl: a case study from theWestern Anatolian Extensional Province
Japan, Nagoya, 26.07.2015 – 02.08.2015, XIX. Inqua Congress

N. Mozafari Amiri N., Ö. Sümer Ö., D. Tikhomirov D.,Ç. Özkaymak Ç., B. Uzel B., S. Ivy-Ochs S., C. Vockenhuber C., H. Sözbilir H., N. Akçar
Holocene destructive seismic periods in Western Anatolia: pace tracking beyond historical data
Switzerland, Basel, 20.-21.11.2015, Swiss Geoscience Meeting

A.M. Müller, M. Döbeli, J. Lachner, M. Suter, H.-A. Synal
Recent gas ionization detector developments at LIP
Austria, Vienna, 15.01.2015, AMS Seminar Vienna

A.M. Müller, M. Döbeli, M. Seiler, H.-A. Synal
Simple Bragg detector for low energy AMS applications
Germany, Heidelberg, 25.03.2015, DPG Spring Meeting

A.M. Müller, M. Döbeli, J. Lachner, M. Suter, H.-A. Synal
Recent gas ionization detector developments at LIP
Spain, Seville, 07.05.2015, AMS Seminar Seville

A.M. Müller, M. Döbeli, H.-A. Synal
Simplified annular gas ionization chamber for backscattering experiments
Croatia, Opatija, 15.06.2015, IBA Conference

A.M. Müller, H.-A. Synal
Accelerator mass spectrometry basics and radiocarbon applications
Germany, Rossendorf, 07.09.2015, EU-Project Sprite Course

A. Neels, X. Maeder, M. Döbeli, A. Dommann, P. Polcik, R. Rachbauer, H. Rudigier, B. Widrig, J. Ramm
Cathodic Arc Evaporation of Oxide Coatings: Investigation of the Phase Transformation at the Target Surface and Deposition of Al and Hf oxides
Croatia, Rovinj, 24.08.2015, European Crystallographic Meeting ECM29

R. Pellitero, B. Rea, M. Spagnolo, J. Bakke, P. Hughes, S. Ivy-Ochs, S. Lukas, H. Renssen, A. Ribolini
A Europe-wide perspective on Younger Dryas glacier-climate
Austria, Vienna, 12.-17.04.2015, EGU General Assembly

C. Schlüchter, N. Akçar, S. Ivy-Ochs, M. Stolz
Relevance of Quaternary alpine paleoglaciations for present day geoengineering projects
Japan, Nagoya, 26.07.2015 – 02.08.2015, XIX. Inqua Congress

S. Schneider, M. Christl, G. Steinhauser, C. Walther
Determination of Plutonium and Uranium in Environmental samples from Fukushima
USA, Kailua-Kona, 17.04.2015, MARC X Conference

S. Schneider, M. Christl, G. Steinhauser, C. Walther
Determination of Plutonium and Uranium in Environmental samples from Fukushima
Germany, Heidelberg, 27.03.2015, DPG Spring Meeting

J. Schoonejans, V. Vanacker, S. Opfergelt, Y. Ameijeiras-Mariño, P. Kubik
Spatial gradient of chemical weathering and its coupling with physical erosion in the soils of the Betic Cordillera (SE Spain)
Austria, Vienna, 14.04.2015, EGU General Assembly

J. Schoonejans, V. Vanacker, N. Bellin, P.W. Kubik, A. Molina, S. Opfergelt, Y. Ameijeiras-Mariño, R. Orteg-Perez
Soil formation and erosion in response to natural and anthropogenic disturbances
Belgium, Louvain-la-Neuve, 31.03.2015, Scientific Symposium, Doctor Honoris Causa UCLouvain

M. Schulte-Borchers, M. Döbeli, A.M. Müller, M. George, H.-A. Synal
Recent progress on the new MeV SIMS setup at ETH Zurich
Croatia, Opatija, 16.06.2015, IBA Conference

P. Sigmund, O. Osmani, A. Schinner, C. Vockenhuber, M. Thöni, J. Jensen, K. Arstila, J. Julin, H. Kettunen, M.I. Laitinen, M. Rossi, T. Sajavaara, H.J. Whitlow
Structure in the velocity dependence of heavy-ion energy-loss straggling
Croatia, Opatija, 15.06.2015, IBA Conference

A. Sookdeo, L. Wacker, S. Fahrni, C. McIntyre, M. Friedrich, F.Reinig, B. Kromer, U. Büntgen
Speed Dating: a rapid way to determine the radiocarbon age of wood by EA-AMS
Senegal, Dakar, 30.11.2015, Radiocarbon Conference

M. Suter
Interaktive Programme für die Modellierung von Beschleunigermassenspektrometrie
Germany, Heidelberg, 25.03.2015, DPG Spring Meeting

H.-A. Synal
The laboratory of Ion Beam Physics
Switzerland, Zurich, 20.02.2015, Scientific Advisory Board Meeting of the Cologne AMS facility

H.-A. Synal
Progress in Accelerator Mass Spectrometry
Croatia, Opatija, 16.06.2015, IBA Conference

H.-A. Synal
Progress in Accelerator Mass Spectrometry
Spain, Seville, 08.10.2015, International Conference on optimization of accelerators, OPAC-2015

H.-A. Synal
Progress in Accelerator Mass Spectrometry
USA, New Orleans, 14.07.2015, Current Trends in Mass spectrometry

H.-A. Synal
Progress in Accelerator Mass Spectrometry
Italy, Benevento, 23.10.2015, 1st International Conference on Metrology for Archaeology

H.-A. Synal
The ETH Zurich Laboratory of Ion Beam Physics
Switzerland, Basel, 12.06.2015, NuPECC Meeting Basel

H.-A. Synal, S. Fahrni, A. Sookdeo, L. Wacker, D. Galvan, T. Knowles, R. Evershed
How far can we get? One permil radiocarbon measurements on a single cathode with a MICADAS instrument
Senegal, Dakar, 30.11.2015, Radiocarbon Conference

C. Terrizzano, R. Zech, E. García Morabito, M. Yamin, N. Haghipour, L. Wüthrich, M. Christl
Neotectonic deformation versus climate control in the Central Andes of Argentina, insights from ^{10}Be Surface Exposure Dating
Austria, Vienna, 14.04.2015, EGU General Assembly

V. Vanacker, J. Schoonejans, N. Bellin, A. Molina, M. Christl
Anthropogenic erosion rates, as a function of human disturbance to vegetation
Germany, Bonn, 18.08.2015, PAGES GLOSS Workshop

V. Vanacker, Y. Ameijeiras-Mariño, N. Bellin, P.W. Kubik, A. Molina, S. Opfergelt, R. Ortega-Perez,
J. Schoonejans
Anthropogenic disturbances to soil systems. New insights from cultural landscapes in the Western Mediterranean
USA, Palo Alto, 15.04.2015, Invited talk, School of Earth, Energy & Environmental Services, Stanford University

V. Vanacker, Y. Ameijeiras-Mariño, N. Bellin, P.W. Kubik, A. Molina, S. Opfergelt, R. Ortega-Perez,
J. Schoonejans
Human disturbances to soil systems in the Western Mediterranean
USA, Denver, 17.03.2015, Invited talk, University of Colorado Denver

S. Vattioni, I. Hajdas, M. Strasser, R. Grischott, T. Sormaz
Lake level reconstruction of lake Sils, Engadine valley
Switzerland, Basel, 21.11.2015, Swiss Geoscience Meeting

C. Vockenhuber, S. Maxeiner, A.M. Müller,M. Suter, H.-A. Synal
Ion Matter Interactions and its relevance for Accelerator Mass Spectrometry
Denmark, Kerteminde, 22.09.2015, Eigth International Meeting on Recent Developments in the Study of Radiation Effects in Matter

C. Vockenhuber, M. Thöni, J. Jensen, K. Arstila, J. Julin, H. Kettunen, M.I. Laitinen, O. Osmani, M. Rossi,
T. Sajavaara, A. Schinner, P. Sigmund, H.J. Whitlow
Energy-loss straggling of MeV heavy-ions in gases
Denmark, Kerteminde, 22.09.2015, Eigth International Meeting on Recent Developments in the Study of Radiation Effects in Matter

C. Vockenhuber, M. Thöni, J. Jensen, K. Arstila, J. Julin, H. Kettunen, M.I. Laitinen, M. Rossi, T. Sajavaara,
H.J. Whitlow
Dedicated experiments for reliable measurements of energy-loss straggling
Croatia, Opatija, 15.06.2015, IBA Conference

C. Vockenhuber
Isobar-separation of intense beams at 6 MV Tandem accelerators
Germany, Heidelberg, 25.03.2015, DPG Spring Meeting

L. Wacker, J. D. Galvan, J. Wunder, U. Büntgen
Extraterrestrial evaluation of global scale tree ring dating in the first millennium CE
Senegal, Dakar, 30.11.2015, Radiocarbon Conference

C. Welte, L. Wacker, B. Hattendorf, M. Christl, J. Koch, D. Günther, H.-A. Synal
^{14}C Analysen von karbonatischen Klimaarchiven mittels Laser Ablation - AMS
Germany, Heidelberg, 25.03.2015, DPG Spring Meeting

C. Welte, B. Hattendorf, L. Wacker, M. Christl, J. Koch, H.-A. Synal, D. Günther
First ^{14}C-Scans on Carbonate Records by Laser Ablation- AMS
Switzerland, Beatenberg, 10.04.2015, Chanalysis 2015

C. Welte, B. Hattendorf, L. Wacker, M. Christl, J. Koch, J. Fohlmeister, S.F.M. Breitenbach, L.F.Robinson, A.H. Andrews, H.-A. Synal, D. Günther
Accessing ^{14}C Profiles in Carbonate Records using Laser Ablation - Accelerator Mass Spectrometry
Czech Republic, Prague, 17.08.2015, Goldschmidt Conference

C. Welte, L. Wacker, B. Hattendorf, M. Christl, J. Fohlmeister, S.F.M. Breitenbach, L.F.Robinson, J.R. Farmer, A.H. Andrews, J. Koch H.-A. Synal, D. Günther
Rapid High Resolution ^{14}C-Analysis of Carbonate Records by Laser Ablation - AMS
Senegal, Dakar, 30.11.2015, Radiocarbon Conference

C. Wirsig, S. Ivy-Ochs, N. Akçar, M. Lupker, K. Hippe, L. Wacker, C. Vockenhuber, H.-A. Synal
Erkenntnisse über die Geschichte eines Alpengletschers durch Kombination von kosmogenem Be-10, in-situ C-14 und Cl-36
Germany, Heidelberg, 25.03.2015, DPG Spring Meeting

C. Wirsig, S. Ivy-Ochs, J. Reitner, M. Christl, C. Vockenhuber, M. Bichler, M. Reindl
Quantifying subglacial erosion rates at Goldbergkees, Hohe Tauern (Austria) with cosmogenic ^{10}Be and ^{36}Cl
Switzerland, Basel, 21.11.2015, Swiss Geoscience Meeting

H. Wittmann, F. v. Blanckenburg, N. Dannhaus, J. Bouchez, J. Gaillardet, J.L. Guyot, L. Maurice, H. Roig, N. Filizola, M. Christl
Denudation and weathering rates from meteoric $^{10}Be/^{9}Be$ ratios in the Amazon basin
Czech Republic, Prague, 18.08.2015, Goldschmidt Conference

L. Wüthrich, R. Zech, N. Haghipour, C. Terrizzano, M. Christl, C. Gnägi, H. Veit, S. Ivy-Ochs
Depth profile dating in the Swiss Midlands: deposition ages versus erosion
Austria, Vienna, 14.04.2015, EGU General Assembly

L. Wüthrich, R. Zech, N. Haghipour, C. Terrizzano, M. Christl, C. Gnägi, H. Veit, S. Ivy-Ochs
^{10}Be depth profile dating in the Swiss Midlands: deposition ages versus erosion
Austria, Vienna, 12.-17.04.2015, EGU General Assembly

S. Yeşilyurt, N. Akçar, U. Doğan, V. Yavuz, S. Ivy-Ochs, C. Vockenhuber, C. Schlüchter
Extensive ice fields in eastern Turkey during the Last Glacial Maximum
Japan, Nagoya, 26.07.2015 – 02.08.2015, XIX. Inqua Congress

C. Zabcı, T. Sançar, D. Tikhomirov, S. Ivy-Ochs, C. Vockenhuber, M. Yazıcı, B.A. Natal'in, H.S. Akyüz, N. Akçar
Understanding the intraplate deformation of the Anatolian Scholle: Insights from the study of the Ovacık Fault (Eastern Turkey)
Austria, Vienna, 12.-17.04.2015, EGU General Assembly

C. Zabcı, T. Sançar, D. Tikhomirov, S. Ivy-Ochs, C. Vockenhuber, N. Akçar
The preliminary slip rates of the Ovacik Fault (Turkey) for the last 16 ka: Implications for the intraplate deformation of the Anatolian scholle
Japan, Nagoya, 26.07.2015 – 02.08.2015, XIX. Inqua Congress

SEMINAR
'CURRENT TOPICS IN ACCELERATOR MASS SPECTRO-
METRY AND RELATED APPLICATIONS'

Spring semester

18.02.2015
Michael Karcher (AWI Bremerhaven), Tracing the Arctic Ocean circulation with ^{129}Iodine

03.03.2015
Victor Alarcon Diez (Université Pierre et Marie Curie UPMC, France), Digital data acquisition with a segmented detector

04.03.2015
Kristina Hippe (ETHZ, Switzerland), Dating metamorphic zircon with U-Pb and Lu-Hf isotope analysis

11.03.2015
Jens Dilling (MPI Heidelberg, Germany), High resolution mass spectrometry of radioactive beams

18.03.2015
Ursula Sojc (ETHZ, Switzerland), Building high-resolution radiocarbon chronologies for the reconstruction of the late Holocene glacier variations and landslide events in the Mont Blanc area, Italy

01.04.2015
Alejandro Ojeda Gonzalez-Posada (PSI, Switzerland), The strong influence of background-gas pressure on the thin film composition

08.04.2015
Jean Nicolas Haas (University of Innsbruck, Austria), First high-resolution dating of a rock glacier world-wide: The Holocene Lazaun rock glacier in the Schnals Valley (South Tyrol, Italy) and its palaeoecological and palaeoclimatic significance

15.04.2015
Giulia Guidobaldi (Pisa University, Italy), Surface Exposure Dating as a tool for reconstructing Late Pleistocene glacial history in Northern Apennines, Italy

22.04.2015
Tessa van der Voort (ETHZ, Switzerland), Radiocarbon: the key to understanding soil organic matter vulnerability

29.04.2015
Jens Leifeld (Agroscope, Switzerland), Separation of old black carbon by thermal methods? A discussion

13.05.2015
Martina Schulte-Borchers (ETHZ, Switzerland), The development of an MeV SIMS

20.05.2015
Roland Purtschert (University of Bern, Switzerland), Cl-36 and Kr-81 dating of groundwater

27.05.2015
Cameron McIntyre (ETHZ, Switzerland), Tracking lab. contamination with ^{14}C swipes

17.06.2015
Selçuk Aksay (ETHZ, Switzerland), A Landform Evolution Investigation with Cosmogenic Nuclide Dating: Sennwald Landslide

Fall semester

26.08.2015
Daniel von Kaenel (ETHZ, Switzerland), Geomorphology and Quaternary Geology of the Bächlital BE

31.08.2015
Louis Lau (University of Lancester, UK), The distribution of I-129 and U-236 off the coast of Japan (NW Pacific Ocean) after the Fukushima accident

16.09.2015
Veit Dausmann (IFM Geomar, Germany), The evolution of climatically driven weathering inputs into the western Arctic Ocean since the late Miocene

23.09.2015
Allen Andrews (NOAA, Washington, USA), Bombs and Fish – How nuclear bombs can tell us about the age of fishes

30.09.2015
Jörg Lippold (University of Bern, Switzerland), Past Ocean Circulation: New insights from Neodymium and 231Pa/230Th isotopes

07.10.2015
Morten Andersen (ETHZ, Switzerland), Uranium cycling on Earth, constraints from isotopic fingerprinting

14.10.2015
Adam Sookdeo (ETHZ, Switzerland), Speed dating - a rapid way of determining the wood's age

21.10.2015
Jakob Schwander (University of Bern, Switzerland), Searching a location to retrieve the longest ice core climate record: RADIX, a new rapid access drilling system

28.10.2015
Pavol Vojtyla (CERN, Switzerland), Radiological environmental aspects of high-power particle accelerators

04.11.2015
Urs Leuzinger (Amt für Archäologie Thurgau, Switzerland), Ein mesolithischer Abri in Muotathal SZ Bisistal-Berglibalm

11.11.2015
Elena Chamizo (CNA, Spain), Status of ^{236}U measurements for oceanography applications at the CNA. ^{236}U from GEOTRACES Equatorial Pacific Zonal Transect (EPZT)

12.11.2015
Christian Kuenz (ETHZ, Switzerland), Quaternary geology and surface exposure dating of erratics on the Albis – Uetliberg – Ridge and Zimmerberg, Zurich

18.11.2015
Guillaume Jouvet (ETHZ, Switzerland), Modelling the trajectory of erratic boulders in the western Alps during the last glacial maximum

25.11.2015
Chiara Uglietti (PSI, Switzerland), The debate on the age of Kilimanjaro's plateau glaciers

02.12.2015
Kevin Kröninger (TU Dortmund, Germany), An introduction to Bayesian Reasoning - Laying the foundation for statistical data analysis

09.12.2015
Tessa van der Voort, Thomas Blattmann (ETHZ, Switzerland), Radiocarbon in Aquatic and Terrestrial Spheres

16.12.2015
Kristina Hippe (ETHZ, Switzerland), Tracing Holocene climate change in the Bolivian Altiplano

THESES (INTERNAL)

Term papers/Bachelor

Bortis Amadé
Microbeam current optimization
ETH Zurich

Jari Klingler
Sedimentological analysis of moraines of different ages and different glacial origin on the crest of the Albis
ETH Zurich

Corinne Singeisen
Morphology and sedimetology of drumlins southwest of Lake Constance and their indications of the Last Glacial Maximum
ETH Zurich

Diploma/Master theses

Selçuk Aksay
The geomorphological evolution of a landscape in a tectonically active region: the Sennwald landslide
ETH Zurich

Silvie Bruggmann
Growth Constraints and Environmental Influences on a Modern Stromatolite, Lagoa Vermelha, Brasil
ETH Zurich

Christian Kuenz
Quaternary geology and surface exposure dating of erratics on the Albis-Uetliberg ridge and Zimmerberg, Zurich
ETH Zurich

Sandro Vattioni
^{14}C analysis of wood from Lake Sils for reconstruction of lake level changes
ETH Zurich

Daniel von Känel
Geomorphology and Quaternary geology of the Bächlital BE
ETH Zurich

Doctoral theses

Merle Gierga
Dating buried molecules-Introducing new applications of small-scale radiocarbon analysis to disentangle the carbon cycle and solve archeological questions
ETH Zurich

Caroline Welte
Laser Ablation coupled with Accelerator Mass Spectrometry for Online Radiocarbon Analysis
ETH Zurich

Christian Wirsig
Constraining the timing of deglaciation of the High Alps and rates of subglacial erosion with cosmogenic nuclides
ETH Zurich

THESES (EXTERNAL)

Diploma/Master theses

Jessica Castleton
The Sentinel rock avalanche of Zion Canyon, Utah
University of Utah (USA)

Christian Eisenach
First ^{10}Be exposure ages from the western Vosges, France, indicating a maximum glacier extent during the Riss glaciation (MIS 6)
University of Münster (Germany)

Caroline Heineke
Surface exposure dating of Late Holocene basalt flows and cinder cones in the Kula volcanic field (Western Turkey) using cosmogenic ^{3}He and ^{10}Be
University of Münster (Germany)

Louis Lau Yik Sze
The distribution of I-129 off the coast of Japan (NW Pacific Ocean) after the Fukushima Accident
Lancaster University (UK)

Melissa Schwab
Coupled Organic and Inorganic Tracers of Particle Flux Processes in the North American Arctic Ocean
ETH Zurich (Switzerland)

Godfroid Thibauld
Evaluation et spatialisation des services écosystemiques en Equateur
UCLouvain (Belgium)

Silavn Wick
Molecular and isotopic constraints on the provenance of organic carbon eroding from Alpine catchments
ETH Zurich (Switzerland)

Doctoral theses

Janine Brunner
Insight into Ca^{2+} Dependent Lipid Scrambling from the Crystal Structure of a TMEM16 Family Member
University of Zurich (Switzerland)

Ruslan Cusnir
In-situ speciation measurements and bioavailability determination of plutonium in natural waters of a karst system using diffusion in thin films (DGT) techniques.
University of Lausanne (Switzerland)

Richard Selwyn Jones
Late Cenozoic behaviour of two Transantarctic Mountain outlet glaciers
Victoria University of Wellington (New Zealand)

Sanjay Mandal Kumar
Topography in passive margins: A case study across the southern Peninsular Indian escarpment using geomorphological analyses, cosmogenic nuclides and low-temperature thermochronometry"
ETH Zurich (Switzerland)

Omar Pecho
Relationships between 3D topology and reaction kinetics in SOFC electrodes
ETH Zurich (Switzerland)

Patrick Reinhard
Alkali-based process and interface engineering of $Cu(In,Ga)Se_2$ thin film solar cells
ETH Zurich (Switzerland)

Mathilde Reinle-Schmitt
The origin and properties of the 2D electron system at polar-oxide interfaces
University of Zurich (Switzerland)

Tino Zimmerling
Electronic Disorder & Charge Trapping in Organic Molecular Semiconductors
ETH Zurich (Switzerland)

Habilitation

Bernadette Hammer-Rotzler
Analysis of the nuclide inventory in MEGAPIE, a proton irradiated lead-bismuth eutectic spallation target
University of Bern (Switzerland)

Tobias Lorenz
Separation and analysis of long-lived radionuclides produced by proton irradiation in lead
University of Bern (Switzerland)

COLLABORATIONS

Australia

Deakin University, Institute for Frontier Materials, Geelong

The Australian National University, Department of Nuclear Physics, Canberra

Austria

AlpS - Zentrum für Naturgefahren- und Riskomanagement GmbH, Geology and Mass Movements, Innsbruck

Geological Survey of Austria, Sediment Geology, Vienna

University of Innsbruck, Institute of Geography, Geology and Botany, Innsbruck

University of Salzburg, Geography and Geology, Salzburg

University of Vienna, VERA, Faculty of Physics, Vienna

Vienna University of Technology, Institute for Geology, Vienna

Belgium

Royal Institute for Cultural Heritage, Brussels

Université catholique de Louvain, Earth and Life Institute, Louvain-la-Neuve

Canada

Chalk River Laboratories, Dosimetry Services, Chalk River

ISOTRACE, Department of Physics, Ottawa

TRIUMF, Vancouver

University of Ottawa, Department of Earth Sciences, Ottawa

China

China Institute for Radiation Protection, Dosimetry Services, Taiyuan city

China Earthquake Administration, Beijing

Chinese Academy of Sciences, Institute of Botany, Beijing

Peking University, Accelerator Mass Spectrometry Lab., Beijing

Denmark

Danfysik A/S, Taastrup

Risø DTU, Risø National Laboratory for Sustainable Energy, Roskilde

Technical University of Denmark, Department of Photonics Engineering, Roskilde

Univ. Southern Denmark, Department of Physics, Chemistry and Pharmacy, Odense

Finnland

University of Jyväskylä, Physics Department, Jyväskylä

France

Aix-Marseille University, Collège de France, Aix-en-Provence

Commissariat à l'énergie atomique et aux énergies alternatives, Laboratoire des Sciences du Climat et de l'Environnement (LSCE), Gif-sur-Yvette Cedex

IRSN, Laboratoire de Radioécologie, Cherbourg-Octeville

Laboratoire de biogeochimie moléculaire, Strasbourg

Université de Savoie, Laboratoire EDYTEM, Le Bourget du Lac

Université Pierre et Marie Curie, Ion Beam Laboratory, Paris

Germany

Alfred Wegener Institute of Polar and Marine Research, Marine Geochemistry, Bremerhaven

BSH Hamburg, Radionuclide Section, Hamburg

Bundesamt für Strahlenschutz, Strahlenschutz und Umwelt, Neuherberg

Deutsches Bergbau Museum, Bochum

GFZ German Research Centre for Geosciences, Earth Surface Geochemistry and Dendrochronology Laboratory, Potsdam

Helmholtz-Zentrum Dresden-Rossendorf, DREAMS, Rossendorf

Helmholtz-Zentrum München, Institut für Strahlenschutz, Neuherberg

Hydroisotop GmbH, Schweitenkirchen

IFM-GEOMAR, Palaeo-Oceanography, Kiel

Leibniz-Institut für Ostseeforschung Warnemünde, Marine Geologie, Rostock

LMU-Munich, Geosciences, Munich

Marum, Micropalaeontology - Paleoceanography and Marine Seimentologie, Bremen

Regierungspräsidium Stuttgart, Landesamt für Denkmalpflege, Esslingen

Reiss-Engelhorn-Museen, Curt-Engelhorn-Zentrum Archäometrie gGmbH, Mannheim

Senckenberg am Meer, Deutsches Zentrum für Marine Biodiversitätsforschung, Wilhelmshaven

University of Cologne, Physics Department and Institute of Geology and Mineralogy, Cologne

University of Hannover, Institute for Radiation Protection and Radioecology, Hannover

University of Heidelberg, Institute of Environmental Physics, Heidelberg

University of Hohenheim, Institute of Botany, Stuttgart

University of Münster, Institute of Geology and Paleontology, Münster

University of Tübingen, Department of Geosciences, Tübingen

Hungary

Hungarian Academy of Science, Institute of Nuclear Research (ATOMKI), Debrecen

India

Inter-University Accelerator Center, Accelerator Division, New Dehli

Italy

CAEN S.p.A., Viareggio

CNR Rome, Institute of Geology, Rome

Geological Survey of the Provincia Autonoma di Trento, Landslide Monitoring, Trento

INGV Istituto Nazionale di Geofisica e Vulcanologia, Sez. Sismologia e Tettonofisica, Rome

University of Bologna, Deptartment Earth Sciences, Bologna

University of Padua, Department of Geosciences, Geology and Geophysics, Padua

University of Pisa, Department of Geology, Pisa

University of Salento, Department of Physics, Lecce

University of Turin, Department of Geology, Turin

Japan

University of Tokai, Department of Marine Biology, Tokai

Liechtenstein

OC Oerlikon AG, Balzers

Oerlikon Surface Solutions AG, Balzers

Mexico

UNAM (Universidad Nacional Autonoma de Mexico), Instituto de Fisica, Mexico

New Zealand

University of Waikato, Radiocarbon Dating Laboratory, Waikato

Victoria University of Wellington, School of Geography, Environment and Earth Sciences, Wellington

Norway

Norwegian University of Science and Technology, Physical Geography, Trondheim

University of Bergen, Department of Earth Science and Biology, Bergen

University of Norway, The Bjerkness Centre for Climate Res., Bergen

Poland

University of Marie Curie Sklodowska, Department of Geography, Lublin

Romania

Horia Hulubei - National Institute for Physics and Nuclear Engineering, Magurele

Singapore

National University of Singapore, Department of Chemistry, Singapore

Slovakia

Comenius University, Faculty of Mathematics, Physics and Infomatics, Bratislava

Slovenia

Geological Survey of Slovenia, Ljubljana

South Korea

KATRI Korea Apparel Testing and Research Institute, Seoul

Spain

Autonomous University of Barcelona, Environmental Science and Technology Institute, Barcelona

University of Murcia, Department of Plant Biology, Murcia

University of Seville, Physics Department and National Center for Accelerators, Seville

Sweden

Lund University, Department of Earth and Ecosystem Sciences, Lund

University of Uppsala, Angström Institute, Upsalla

Switzerland

ABB Ltd, Baden

ABB Ltd, Lenzburg

Amt für Kultur Kanton Graubünden, Archäologischer Dienst, Chur

Centre Hospitalier Universitaire Vaudois, Institut de radiophysique, Lausanne

Dendrolabor Wallis, Brig

Empa, Research Groups: Nanoscale Materials Science, Mechanics of Materials and Nanostructures, X-ray Analytics, Functional Polymers and Hochleistungskeramik, Dübendorf

ENSI, Brugg

ETH Zurich, Departments of: Metals Research and Polymers MATL, Inorganic Chemistry CHAB, Electronics Laboratory ITET, Particle Physics PHYS, Trace Element and Micro Analysis CHAB, and of Health Sciences and Technology; Istitutes of Engineering Geology, Isotope Geoche, Zurich

FAEI, Geneva

Federal Office for Civil Protection, Spiez Laboratory, Spiez

Glas Trösch AG, Bützberg

Gübelin Gem Lab Ltd. (GGL), Luzern

Haute Ecole ARC, IONLAB, La-Chaux-de-Fonds

Helmut Fischer AG, Hünenberg

II-VI Laserenterprise, Zurich

Kanton Bern, Achäologischer Dienst, Bern

Kanton Graubünden, Kantonsarchäologie, Chur

Kanton Solothurn, Kantonsarchäologie, Solothurn

Kanton St. Gallen, Kantonsarchäologie, St. Gallen

Kanton Turgau, Kantonsarchäologie, Frauenfeld

Kanton Zug, Kantonsarchäologie, Zug

Kanton Zürich, Kantonsarchäologie, Dübendorf

Labor für quartäre Hölzer, Affoltern a. Albis

Laboratiore Romand de Dendrochronologie, Moudon

Landesmuseum, Zurich

Office et Musée d'Archéologie Neuchatel, Neuchatel

Paul Scherrer Institut (PSI), Laboratories for Micro and Nanotechnology, for Atmospheric Chemistry, for Radiochemistry and Environmental Chemistry, Materials Group, Radiochemistry, Villigen

Research Station Agroscope Reckenholz-Tänikon ART, Air Pollution / Climate Group, Zurich

Stadt Zürich, Amt für Städtebau, Zurich

Swiss Federal Institute for Forest, Snow and Landscape Reseach (WSL), Landscape Dynamics, Dendroecology and Soil Sciences, Birmensdorf

Swiss Federal Institute of Aquatic Science and Technology (Eawag), SURF, Dübendorf

Swiss Gemmological Institute, SSEF, Basel

Swiss Institute for Art Research, SIK ISEA, Zurich

University of Basel, Departement Altertumswissenschaften und Institut für Prähistorische und Naturwissenschaftliche Archäologie (IPNA), Basel

University of Bern, Department of Chemie and Biochemistry, Climate and Environmental Physics, Oeschger Center for Climate Research and Institute of Geology, and Geography, Bern

University of Freiburg, Faculty of Environmentat and Natural Resources, Freiburg

University of Fribourg, Department of Physics, Fribourg

University of Geneva, Department of Anthropology and Ecology, Geology and Paleontology, and Quantum Matter Physics, Geneva

University of Lausanne, Department of Geology, Lausanne

University of Neuchatel, Department of Geology, Neuchatel

University of Zurich, Department of Geography, Institute of Geography, Abteilung Ur- und Frühgeschichte, Zurich

Turkey

Istanbul Technical University, Faculty of Mines, Istanbul

Tübitak, Marmara Arastirma Merkezi, Gebze Kocaeli

United Kingdom

Brithish Arctic Survey, Cambridge

Durham University, Department of Geography, Durham

Newcastle University, School for History, Classics and Archaeology, Newcastle

Northumbria University, Department of Geography, Newcastle

University of Bristol, School of Chemistry and School of Earth Sciences, Bristol

University of Oxford, Department of Earth Sciences, Oxford

USA

Colorado State University, Department of Environmental and Radiological Health Sciences, Fort Collins

Columbia University, LDEO, New York

Eckert & Ziegler Vitalea Science, AMS Laboratory, Davis

Florida State University, Earth, Ocean & Atmospheric Science, Tallahassee

Harvard University, Engineering, Cambridge

Idaho National Laboratory, National and Homeland Security, Idaho Falls

Lamont-Doherty Earth Observatory, Department of Geochemistry, Palisades

NOAA Fischeries, Pacific Islands Fisheries Science Center, Honolulu

University of Utah, Geology and Geophysics, Salt Lake City

Woods Hole Oceanographic Institution, Center for Marine and Environmental Radioactivity, and Marine Chemistry and Geochemistry, Woods Hole

VISITORS AT THE LABORATORY

Silvana Martin
University of Padua, Department of Geosciences, Padua, Italy
12.01.2015 - 15.01.2015

Victor Alarcon Diez
Université Pierre et Marie Curie UPMC, Institut des NanoSciences de Paris, Paris, France
09.02.2015 - 07.03.2015

Michael Karcher
AWI Bremerhaven, Bremerhaven, Germany
17.02.2015 - 18.02.2015

Ian Vickridge
Université Pierre et Marie Curie UPMC, Paris, France
02.03.2015 - 07.03.2015

Silvana Martin
University of Padua, Department of Geosciences, Padua, Italy
23.03.2015 - 27.03.2015

Valentin Stoytschew
University of Zagreb, Ruder Boskovic Institute, Zagreb, Croatia
30.03.2015 - 11.04.2015

Giulia Guidobaldi
University of Pisa, Earth Sciences Department, Pisa, Italy
01.04.2015 - 31.07.2015

Jakub Zeman
University of Bratislava, Faculty of Mathematics, Physics and Informatics, Dept. Of Nuclear Physics nd
Biophysics, Bratislava, Slovakei
01.04.2015 - 31.07.2015

Stefano Casale
University of Pisa and University of Florence, Earth Sciences Department, Pisa, Florence, Italy
01.04.2015 - 17.07.2015

Ronny Friedrich
Curt-Engelhorn-Zentrum Archäometrie gGmbH, Mannheim, Germany
08.04.2015 - 16.07.2015

Juan Diego Galvan
Swiss Federal Institute for Forest, Snow and Landscape Research (WSL), Birmensdorf, Switzerland
15.04.2015 - 31.12.2015

Tiberiu Bogdan Sava
Department of Tandem Accelerators - IFIN-HH, Magurele-Ilfov, Romania
01.06.2015 - 10.06.2015

Niklaus Burri
Kantonsschule im Lee, Turbenthal, Switzerland
01.06.2015 - 05.06.2015

Leo Klarner
Deutsche Schule Moskau, Moscow, Russia
01.06.2015 - 05.06.2015

Louis Lau
Lancaster University, Lancaster Environmental Center, Lancaster, UK
07.06.2015 - 31.08.2015

Julia Immoor
University of Kiel, Geology, Kiel, Germany
23.06.2015

Christiane Yeman
Unversity of Heidelberg, Environmental Physics, Heidelberg, Germany
23.06.2015

Ewelina Opyrchal
AGH University of Science and Technology, Department of Environmental Analysis, Cartography and
Economic Geology, Krakow, Poland
01.07.2015 - 31.07.2015

Noel Schürch
Gymnasium Thun, Standort Schadau, Thun, Switzerland
10.07.2015

Seung Jae Lee
Korea Institute of Science and Technology KIST, Saarbrücken, Germany
20.07.2015 - 20.09.2015

Katharina Glückel
Helmholtz Zentrum, Munich, Germany
29.07.2015

Monika Isler (Bachelor thesis)
University of Zurich, Institut für Archäologie/FB Prähistorische Archäologie, Zurich, Switzerland
15.09.2015 - 31.12.2015

Amir Sindelar (Bachelor thesis)
University of Zurich, Institut für Archäologie/FB Prähistorische Archäologie, Zurich, Switzerland
15.09.2015 - 31.12.2015

Allen Andrews
National Oceanic and Atmospheric Administration, Washington, USA
21.09.2015 - 25.09.2015

Tim Knowles
University of Bristol, Bristol, UK
28.09.2015 - 30.09.2015

Stefano Casale
University of Pisa and University of Florence, Earth Sciences Department, Pisa, Florence, Italy
01.10.2015 - 31.12.2015

Elena Chamizo
Centro Nacional de Aceleradores (CNA), Seville, Spain
09.11.2015 - 13.11.2015

Ewelina Opyrchal
AGH University of Science and Technology, Department of Environmental Analysis, Cartography and Economic Geology, Krakow, Poland
13.11.2015 - 16.12.2015

Alexander Gogas
Kantonschule Olten, Olten, Switzeraland
16.11.2015 - 20.11.2015

Francesca Quinto
Karlsruher Institut für Technologie, Institut für Nukleare Entsorgung, Eggenstein-Leopoldshafen, Germany
27.11.2015